THE LANGUAGE OF CHEMISTRY

BY 'REMI TIJANI

COPYRIGHT © Habeeb Aderemi Tijani - Asoroda

All rights reserved.

Published in Nigeria by
THABAS PUBLISHING COMPANY

PLOT 28, BLOCK X,

FEDERAL SITE AND SERVICES SCHEME,

APAPA, MONIYA,

IBADAN.

NIGERIA.

Tel. +2348037146269

+2348097502071

E-mail:- thabaspublishers@yahoo.com

asoroda63@yahoo.com

First impression 2009

Kindle Edition 2014

ISBN: 1493671766 (CreateSpace)

ISBN-13: 978-1493671762 (CreateSpace)

DEDICATION

This work is dedicated:

To The Almighty God, my source of wisdom.

To my parents, my sources of life.

And

To my brothers and sisters in Chemistry all over the world.

'Remi Tijani

(September 2013)

REFLECTION

"I am a hidden treasure

I wish to be known

So

I created the Worlds".

(HADITH QUDSI)

TABLE OF CONTENTS

Copyright Page *2*
Dedication *3*
Reflection *4*
Table of Contents *5-8*
Foreword *9*
Preface *10*
Acknowledgement *11*

CHAPTER ONE: ATOM, THE BUILDING BLOCKS OF MATTER *12*

STRUCTURE OF THE ATOM *13*
PROPERTIES OF ATOMS *15*
ATOMIC MASS AND WEIGHT *15*

CHAPTER TWO: THE MOLE CONCEPTRELATIVE ATOMIC MASSES *17*

RELATIVE MOLECULAR MASS *19*
THE CONCEPT OF MOLE *20*
MOLAR VOLUME *22*
Practice questions for Students C2 *22*

CHAPTER THREE: THE CONCEPT OF OXIDATION NUMBER *25*

INTRODUCTION *25*
DEFINITION *26*
IONIC BONDING and COVALENT BONDING *27*
IONIC BONDING *27*
COVALENT BONDING *28*
OXIDATION NUMBER *29*
Practice questions for Students C3 *31*

CHAPTER FOUR: CHEMICAL NOMENCLATURE *32*

IONS *33*
CATIONS *33*
Mono atomic Cations *33*
Polyatomic cations *34*
The Hydroxonium ion H3O+ *35*
ANIONS *35*
Polyatomic anions *35*
OXIDES *37*
Acidic oxides *37*
Basic Oxides *38*
Amphoteric Oxides *38*
Neutral Oxides *39*
Peroxides *39*
ACIDS *40*

BASES *40*
SALTS (inorganic) *41*
CLASSIFICATION OF SALTS *42*
Acid Salts *42*
Normal Salts *43*
Basic Salts *43*
Hydrated Salt *43*
HALIDES *44*
Practice questions for the Students. C4 *45*

CHAPTER FIVE: THE SPEAKABLE LANGUAGE *46*
CONSERVATION LAWS *43*
REACTIONS IN SOLUTION (conservation law continues) *44*
COMPLEX EQUATIONS *45*
Reactions Taking Place in Acid Medium *48*
Reactions taking place in basic medium *48*
Practice questions for Students. C5 *51*

CHAPTER SIX: ELECTROLYSIS: ILLUSION OR REALITY? *52*
PROCESS OF ELECTROLYSIS *53*
Laws of Electrolysis *55*
Mechanism of Conduction in Electrolysis *58*
Applications (Uses) Of Electrolysis *59*
Effects or Applications of Electroplating *62*
Practice questions for the Students. C6 *62*

CHAPTER SEVEN: DEFINITIONS OF SOME CONCEPTS & REACTION PROCESSES *63*
A COMPOUND *63*
DECOMPOSITION REACTION *63*
DEHYDROGENATION REACTION *63*
DISPLACEMENT REACTION *63*
ELECTROCHEMICAL SERIES *63*
DISPROPORTIONATION REACTION *64*
ELECTRON *65*
ELEMENTS *66*
ELECTRO NEGATIVITY *65*
HYDROGENATION REACTION *65*
HYDROLYSIS (HYDROXIDATION) *65*
ION EXCHANGE *66*
IONIZATION *66*
NEUTRALIZATION REACTIONS *66*
OXYGENATION REACTIONS *66*
REVERSIBLE REACTION *66*
SUBSTANCE *66*

SUBSTITUTION REACTIONS **66**
THE MOLE **67**
MOLARITY **67**
MOLALITY **67**
NORMALITY **67**
Law of Conservation of Mass [Lavoisier 1794] **68**
Law of Constant Composition or Definite Proportion [Proust 1779] **68**
Law of Multiple Proportions [John Dalton 1803] **68**
The Law of Equivalent **69**
Ionic Theory **69**
Gay-Lussac's Law of Combine Volume [1808] **69**
Avogadro's Hypothesis [1811] **69**
Boyles' Law **70**
Charles' Law **71**
Pressure Law **71**
Dalton's Law of Partial Pressure [1810] **72**
Graham's Law of Diffusion **72**
Le Chatelier's Principle [1885] **73**
Factors Affecting the Rate of Chemical Reaction **74**
Properties of a Catalyst **74**
Free Energy change ΔG **75**
Entropy change of a reaction ΔS **75**
Enthalpy change ΔH **75**
Heat of formation **75**
Exothermic reaction **75**
Endothermic reaction **75**
Thermoneutral reactions **76**
CONSTANTS **77**
Practice questions for Students C7 **78**
A Table of Oxidation Number of some common elements **79**
CHEMICAL ELEMENTS SORTED IN ALPHABETICAL ORDER WITH SOME FACTS ABOUT THEIR DISCOVERY **82**
CHEMICAL ELEMENTS SORTED IN ALPHABETICAL ORDER WITH SOME INTERESTING FACTS ABOUT THEM **88**
CHECK LISTS OF CHEMICAL FORMULAS AND THEIR IUPAC NOMENCLATURE **91**
A CHECK LIST OF BALANCED CHEMICAL EQUATIONS **99**
DECOMPOSITION REACTION **99**
ADDITION REACTIONS **100**
DEHYDROLIZATION (DEHYDRATION) REACTION **101**
DEOXYGENATION REACTION **102**
DISPLACEMENT REACTION **103**
ELECTRON TRANSFER **105**
HYDROLIZATION REACTIONS (HYDROLYSIS) **107**
HYDROGENATION REACTIONS **109**

ION EXCHANGE REACTION *109*
IONIZATION REACTIONS *109*
NEUTRALIZATION REACTIONS *110*
OXYGENATION REACTION *112*
REDOX REACTIONS *113*
REVERSIBLE REACTION *115*
SUBSTITUTION REACTIONS *116*
NUCLEAR REACTIONS *117*
REFERENCES *118*
ABOUT THE AUTHOR *118*

FOREWORD

This book, "the Language of Chemistry", is self-introducing as Chemistry comes more often in symbols and structures. Chemical symbols and structures are the language of Chemistry. This is akin to what scientists' say of Mathematics; that it is the language of science in symbolic forms. The author tries to present Chemistry in a simple way, as a web of ideas. I will, therefore, want to see this as a book of ideas; the strong and stimulating ideas which underlie Chemistry in this era when there is dwindling interest among students to study science, particularly Chemistry.

All ideas in Chemistry are more practical than theoretical. Even when a chemical idea is presented in a theoretical way, the end result is in the laboratory where ideas are converted to reality. In the same way, the language of Chemistry in this book is the language of the laboratory. The dwindling fortune in the sciences in Nigeria is because less emphasis is laid on practical aspects when compared to the theoretical aspects. The usefulness of a book, which contains scientific ideas, can only be judged by results.

I recommend this book for both beginners and those who have had some experience in Chemistry. To the beginner, the book can delight the mind. To the experienced, it can fire the mind. The ideas in this book are, therefore, not just for people wanting to pass one examination or the other, but for those who want to use Chemistry as a career and a hobby. I do see people write "reading" as a hobby but what they read is often left back-stage. This book offers the opportunity to include the "Language of Chemistry" in that list.

The book is timely. It has come when there is economic meltdown all over the world and as such finances for importation of books may not be available. It has come at a point when there is emphasis on the teaching and learning of science at all levels to catalyse development. I recommend the book to all those who do not mind the "English of Chemistry" but the language of it.

Professor M. U. Adikwu, FAS

Nsukka/Abuja, 2013

PREFACE

As a little boy, I used to wonder about the being of human race. Do they not say that man evolved from Ape? How come Ape is still Ape? NO! Man was never Ape! Man is a unique creature of God! But, what has God put in man that makes him so unique that he has been able to create his world? Knowledge and wisdom! I realized. Man has created and recreated his world. He created his comfort, created his destruction and perhaps, he has created his ultimate extinction. He calls it the world of science. Yes! I call it the world of Chemistry. Why not? Is it not Chemistry that does it all?

The day I realized these facts, I prayed to God. "Oh God of wisdom, make me one of the preachers of the GOSPEL OF CHEMISTRY."

In 1989 when I started the work on 'HAT UI 89, A CHEMISTRY IN DOOR GAME'; I did not know that I was creating a world, until I consulted with Professors: S. T. Bajah, J. O. Akinboye and J. I. Okogun all from the University of Ibadan. They urged me to perfect this new world so that the generations to come would dwell in it.

I accepted the challenge and set to work, which lasted two decades. Now I thought it worthwhile to write this hand-book which I believe would serve as an invaluable companion to the present and future generations of scientists in the world of Chemistry.

In this small book, I have tried to remove some hidden barriers in order to reach proper understanding of some important concepts in the Subject; such concept like oxidation number, mole concept, IUPAC nomenclature, writing and balancing of chemical equations etc. Hence, the book is called; The Language of Chemistry.

In it also is a list of about one thousand balanced chemical equations as well as an equal number of chemical formulas with their IUPAC nomenclatures to assist the players of 'HAT: THE-CHEMISTRY GAME' in their adventures. It also contain names of chemical elements with their properties, date of discovery, name of discoverer and some interesting facts about them.

This book and the game would be most useful to the following categories of people:

1. The science teachers who wish to keep abreast with developments in the field.
2. The science students at Senior Secondary Schools and higher institutions.
3. Researchers in Chemistry and related fields.
4. Those who believe in being practical.

Though I have double-checked all the facts in this book, I hope that those who use it will be frank in calling to my attention any error(s) or omission(s) they may note.

'Remi Tijani
(September 2013)

ACKNOWLEDGEMENT

This work would not be complete if I do not acknowledge the authors of the various Chemistry books and journals I have read in my efforts to learn Chemistry. Though I cannot remember all their names, they have all contributed to making me confident and comfortable in the world of Chemistry.

In the list of my motivators are Prof. J. I. Okogun of Chemistry Department University of Ibadan; Prof. J. O. Akinboye of Guardian & Counseling Dept., University of Ibadan; Prof. Sam Tunde Bajah of Institute of Education, University of Ibadan; Dr. Uzo Egbugara and Prof. G. M. O. Onwu of Teacher Education, University of Ibadan; They are my channels of inspiration.

I acknowledge with thanks, the efforts of my friend and sister, Dr (Mrs) Alima Alabi of the Department of Chemistry, University of Ibadan; who reviewed this edition of the book at a very short notice.

Also Prof. D.O.S. Noibi of the Department of Arabic and Islamic Studies, University of Ibadan; for his interest in my work and the encouragement of the muslim community of University of Ibadan. My thanks also go to Alhaji K. I. Abubakar, the Chief Lab. Technologist, Federal University of Technology, Owerri, who assisted me in experimentation at the formative level of this book. Mr V. O. Masagbor, former chief librarian of Ahmadu Bello University, Zaria who assisted me with journals.

I Specially, thank my mentor, professor Michael U. Adikwu (FAS), the national co-ordinator of STEP-B (Science and Technology Education Post Basic) projects in Nigeria, whose interest and encouragement in my work has mid-wifed the birth of this book and the Chemistry game (HAT: The chemistry game) after a long period of conception.

I will never forget surveyor Oshibeluwo, the then O/C Federal Survey Office, Owerri and Mr. Adebayo Olaloye Isaac of NBC (Coca Cola) Owerri who were more of brothers to me than friends during my NYSC year in Owerri many years ago. Also Engr. Yinus Gbadamosi of Himus Petroleum who has always been like a brother to me.

Finally, my thanks go to my parents, Alhaji Tijani Adedigba Asoroda of blessed memory and Alhaja Sikirat Abewon Tijani Asoroda for sending me to school, my uncle like a father, Alhaji Ibrahim Ashiru, my wife, my children, my brothers, sisters and friends for their patience with me when I had no time to please them. My well-wishers and everybody who have encouraged or discouraged me at one time or the other in my life. I leave you all to Almighty God for your rewards.

'Remi Tijani
(December 2013)

CHAPTER ONE

ATOM !

THE BUILDING BLOCKS OF MATTER

Atom is the tiny basic building block of matter. All the materials on earth are composed of various combinations of atoms. Atoms are the smallest particles of a chemical element that still exhibit all the chemical properties unique to that element. It is so tiny that a row of 100 million atoms would be only about a centimeter long!

Understanding atoms is the key to understanding the physical world. More than 100 different elements exist in nature, each with its own unique atomic makeup. The atoms of these elements react with one another and combine in different ways to form virtually unlimited number of chemical compounds. When two or more atoms combine, they form a molecule. For example, two atoms of the element hydrogen (abbreviated H) combine with one atom of the element oxygen (O) to form a molecule of water (H_2O).

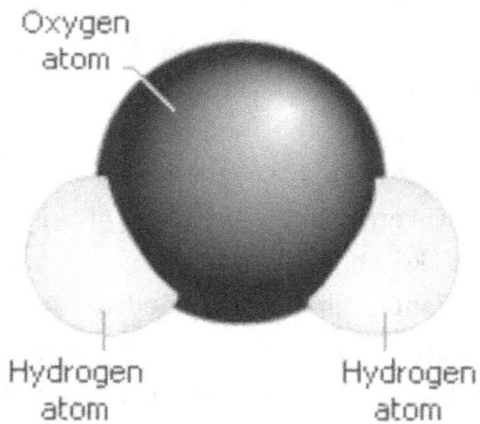

Since all matter, from their formation in the early universe to present-day biological systems; consists of atoms, understanding their structures and properties play a vital role in Physics, Chemistry, and Medicine. In fact, knowledge of atoms is essential to the modern scientific understanding of the complex systems that govern the physical and biological worlds. Atoms and the compounds they form play a part in almost all processes that occur on earth and in space.

All organisms rely on a set of chemical compounds and chemical reactions to digest food, transport energy, and reproduce. Stars such as the sun rely on reactions in atomic nuclei to produce energy. Scientists duplicate these reactions in laboratories on earth and study them to learn about processes that occur throughout the universe.

Throughout history, people have sought to explain the world in terms of its most basic parts. Ancient Greek philosophers conceived the idea of the atom, which they defined as the smallest possible piece of a substance. The word atom comes from the Greek word meaning "not divisible." The ancient Greeks also believed this fundamental particle was indestructible. Scientists have since learned that atoms are not indivisible but made of smaller particles, and atoms of different elements contain different numbers of each type of these smaller particles.

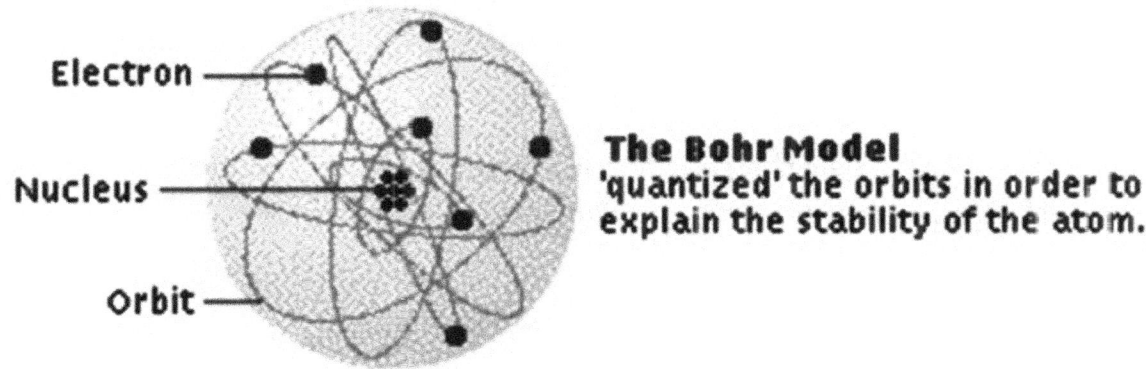

STRUCTURE OF THE ATOM

Atoms are made of smaller particles, called electrons, protons, and neutrons. An atom consists of a cloud of electrons surrounding a small, dense nucleus of protons and neutrons. Electrons and protons have a property called electric charge, which affects the way they interact with each other and with other electrically charged particles. Electrons carry a negative electric charge, while protons carry a positive electric charge. The negative charge is the opposite of the positive charge, and, like the opposite poles of a magnet, these opposite electric charges attract one another. Conversely, like charges (negative and negative, or positive and positive) repel one another. The attraction between an atom's electrons and its protons holds the atom together. Normally, an atom is electrically neutral, which means that the negative charge of its electrons is exactly equaled by the positive charge of its protons.

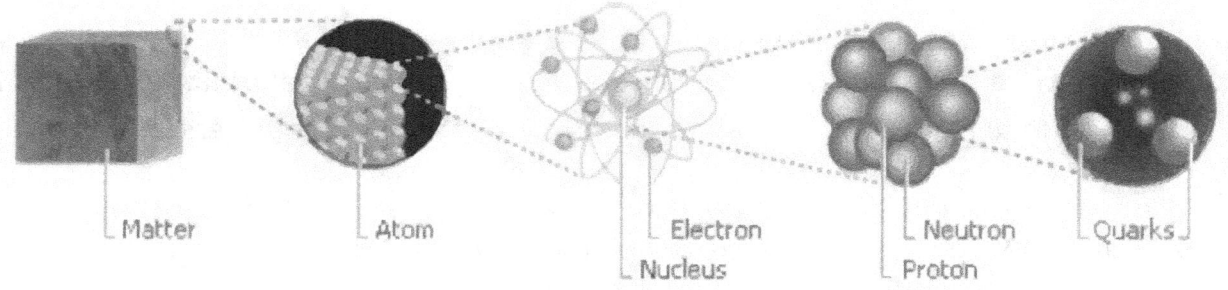

The nucleus contains nearly all of the mass of the atom, but it occupies only a tiny fraction of the space inside the atom. The diameter of a typical nucleus is only about 1×10^{-14} m, or about 1/100,000 of the diameter of the entire atom. The electron cloud makes up the rest of the atom's overall size. If an atom were magnified until it was as large as a football stadium, the nucleus would be about the size of a grape placed in its Centre!

The Electrons

Electrons are tiny, negatively charged particles that form a cloud around the nucleus of an atom. Each electron carries a single fundamental unit of negative electric charge, or −1 (e^-).

The electron is one of the lightest particles with a known mass. A droplet of water weighs about a billion, billion, billion times more than an electron. Scientists believe that electrons are one of the fundamental particles of Physics, which means they cannot be split into anything smaller. Scientists also believe that electrons do not have any real size, but are instead, true points in space—that is, an electron has a radius of zero.

Electrons act differently than everyday objects because electrons can behave as both particles and waves. Actually, all objects have this property, but the wavelike behavior of larger objects, such as sand, marbles, or even people, is too small to measure. In very small particles, wave behavior is measurable and important. Electrons travel around the Nucleus of an atom, but because they behave like waves, they do not follow a specific path like a planet orbiting the sun does. Instead, they form regions of negative electric charge around the nucleus. These regions are called orbitals, and they correspond to the space in which the electron is most likely to be found. As we will discuss in some later chapters, orbital have different sizes and shapes, depending on the energy of the electrons occupying them.

Protons and Neutrons

Protons carry a positive charge of +1, exactly the opposite electric charge as electrons. The number of protons in the nucleus determines the total quantity of positive charge in the atom. In an electrically neutral atom, the number of the protons and the number of electrons are equal, so that the positive and negative charges balance out to zero. The proton is very small, but it is fairly massive compared to the other particles that make up atom. A proton's mass is about 1,840 times the mass of an electron.

In 1932, British physicist, Sir James Chadwick, discovered the subatomic particle known as the neutron, filling in a key missing piece in Science's understanding of the atom. The neutron has no electric charge, but has a mass nearly equivalent to that of a proton, and is a component of the nucleus of an atom. Chadwick received the 1935 Nobel Prize in Physics for his discovery.

Neutrons are about the same size as protons but their mass is slightly greater. Without neutrons present, the repulsion among the positively charged protons would cause the nucleus to fly apart. Consider the element helium, which has two protons in its nucleus. If the nucleus did not contain neutrons as well, it would be unstable because of the electrical repulsion between the protons. A helium nucleus needs either one or two neutrons to be stable. Most atoms are stable and exist for a long period of time, but some atoms are unstable and spontaneously break apart and change, or decay, into other atoms.

Unlike electrons, which are fundamental particles, protons and neutrons are made up of other smaller particles called *quarks*.

PROPERTIES OF ATOMS

Atoms have several properties that help distinguish one type of atom from another and determine how atoms change under certain conditions.

Atomic Number

Each element has a unique number of protons in its atoms. This number is called the atomic number (abbreviated Z). Because atoms are normally electrically neutral, the atomic number also specifies how many electrons an atom will have. The number of electrons, in turn, determines many of the chemical and physical properties of the atom. The lightest atom, hydrogen, has an atomic number equal to one, contains one proton, and (if electrically neutral,) one electron. The most massive stable atom found in nature is bismuth (Z = 83). More massive unstable atoms also exist in nature, but they break apart and change into other atoms over time. Scientists have produced even more massive unstable elements in laboratories.

Mass Number

The total number of protons and neutrons in the nucleus of an atom is the mass number of the atom (abbreviated A). The mass number of an atom is an approximation of the mass of the atom. The electrons contribute very little mass to the atom, so, they are not included in the mass number. A stable helium atom can have a mass number equal to three (two protons plus one neutron) or equal to four (two protons plus two neutrons). Bismuth, with 83 protons, requires 126 neutrons for stability. So its mass number is 209 (83 protons plus 126 neutrons).

ATOMIC MASS AND WEIGHT

Scientists usually measure the mass of an atom in terms of a unit called the atomic mass unit (abbreviated a.m.u). They define an a.m.u as exactly 1/12 the mass of an atom of carbon with six protons and six neutrons. On this scale, the mass of a proton is 1.00728 a.m.u and the mass of a neutron is 1.00866 a.m.u. The mass of an atom measured in a.m.u is nearly equal to its mass number.

Scientists can use a device called mass spectrometer to measure atomic masses. A mass spectrometer removes one or more electrons from an atom. The electrons are so light that removing them hardly changes the mass of the atom at all. The spectrometer then sends the atom through a magnetic field (a region of space that exerts a force on magnetic or electrically charged particles). Because of the missing electrons, the atom has more protons than electrons and hence, a net positive charge. The magnetic field bends the path of the positively charged atom as it moves through the field. The amount of bending depends on the atom's mass. Lighter atoms will be affected more strongly than heavier atoms. By measuring how much the atom's path curves, a scientist can determine the atom's mass.

The atomic mass of an atom, which depends on the number of protons and neutrons present, also relates to the atomic weight of an element. Weight usually refers to the force of gravity on an object, but atomic weight is really just another way to express mass. An element's atomic weight is given in grams. It represents the mass of one mole (6.02×10^{23} atoms) of that element. Numerically, the atomic weight and the atomic mass of an element are the same, but atomic weight is expressed in grams while atomic mass is in atomic mass units (a.m.u). So, the atomic weight of hydrogen is 1 gram and the atomic mass of hydrogen is 1 a.m.u.

CHAPTER TWO

THE MOLE CONCEPT

INTRODUCTION

Mole is one of the most important units of measurement in Science. But, for the development of this important concept, there would have been little or no progress in the world of quantitative analysis in Chemistry since it is this concept that makes provision for measuring the amount of substance present in a given mass. The concept thus extended the questions that were formerly common with the chemists from: "What happens? How did it happen?" to "What quantity (amount) of each material was involved when it happened?"

RELATIVE ATOMIC MASSES

Until 1920 when Mass Spectrometer was invented, the chemists had almost given up the thought of obtaining the accurate mass of individual atoms. This was because scientists at that time knew that atoms were so small, and thinking of obtaining their accurate masses was like thinking of the impossible. What preoccupied their minds was how to obtain the collective mass of a group of atoms and how the masses of atoms of different elements relate to one another. For example, 'which is heavier, an oxygen atom or a nitrogen atom'? Or is an atom of oxygen heavier than an atom of lithium? If yes, by how many times? Such thoughts led to the formation of the Relative Atomic Mass concept.

For instance,

100 atoms of Hydrogen weigh 100 mass units

100 atoms of Oxygen weighs 1600 mass unit

100 atoms of Nitrogen weighs 1400 mass unit

100 atoms of Lithium weigh 700 mass units

100 atoms of Sodium weighs 2300 mass unit.

Under this concept, masses of atoms of all elements were compared with the mass of a standard atom. The standard introduced then was an atom of hydrogen, whose mass was fixed at unity [1]. Thus, it was expected that this comparison would give whole numbers greater than unity for other elements. At that time, the Relative Atomic Mass was defined in terms of hydrogen as:

The mass of one atom of an element compared with the mass of one atom of hydrogen.

Mathematically,

$$\frac{\text{Mass of one atom of any element}}{\text{Mass of one atom of Hydrogen}} = \text{Relative atomic mass } A_r \text{, of that element}$$

Example :

If the mass of an atom of Zinc is X, what is its relative atomic mass?

$$A_r = \frac{X}{1} = X$$

Thus, when you are told that the relative atomic mass of nitrogen is 14 (^{14}N) or that of sodium is 23 (^{23}Na), what you are being told is that every atom of nitrogen has a mass which is fourteen (14) times heavier than one atom of hydrogen, and every atom of sodium is 23 times heavier than an atom of hydrogen. Note that by this comparison of elements with hydrogen, we are also comparing the masses of other elements with one another. From the above example, nitrogen is 14 and sodium is 23, it therefore implies that an atom of sodium is heavier than that of nitrogen in the proportion of 23 to 14.

When the presence of isotopes was discovered in the late 1920s, scientists were forced to drop the use of hydrogen atom as the standard for comparing atomic masses since hydrogen has three isotopes:

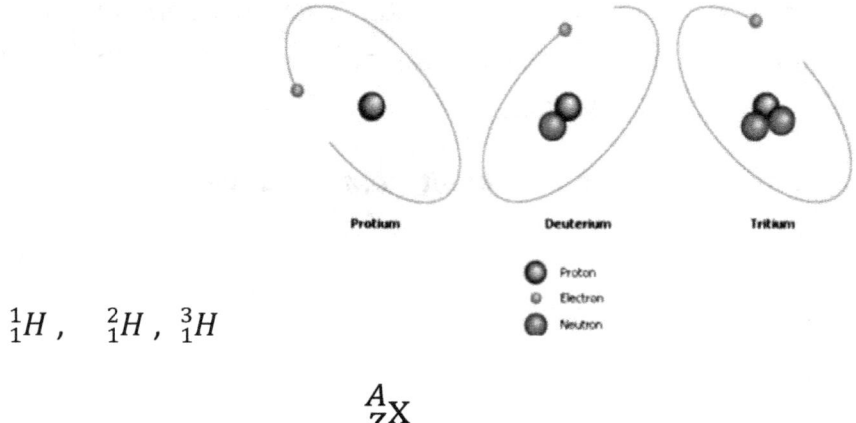

^1_1H , ^2_1H , ^3_1H

$$^A_ZX$$

Where X is any atom, Z is its atomic number and A is its mass number.

This led to the adoption of a new standard; the carbon 12 isotope; which gives a more accurate result.

(Carbon has three naturally occurring isotopes: carbon-12, carbon-13 and carbon-14. However, carbon-12 constitutes 98.89 percent of all carbon atoms and serves as the standard for the atomic mass scale; carbon-13 is the only magnetic isotope, which makes it very important for structural studies of compounds containing carbon. Carbon-14 is produced by cosmic ray bombardment of nitrogen. It is radioactive with a half-life of 5,760 years).

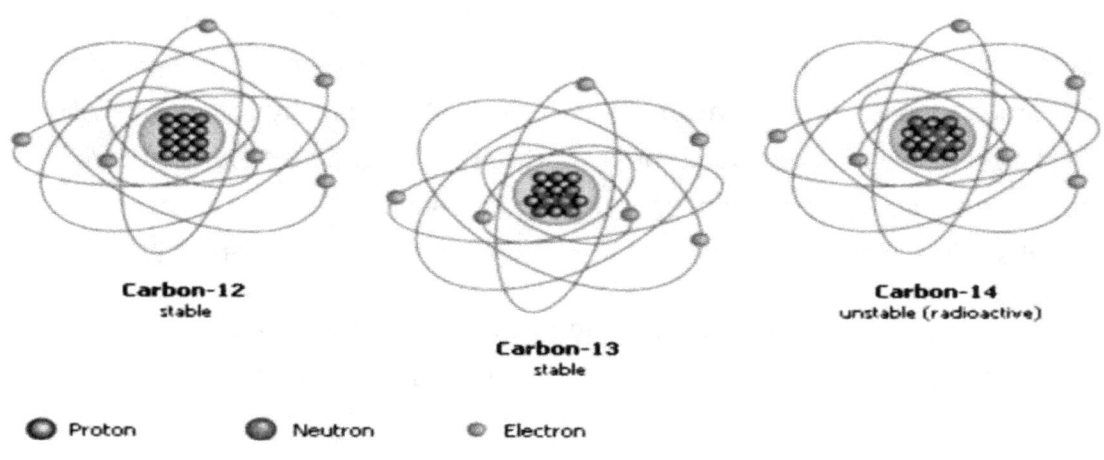

The relative atomic mass has since then been redefined thus:

The relative atomic mass of an element is the number of times which the mass of one atom of that element is heavier than one-twelfth (1/12) of the mass of one atom of Carbon-12 ($^{12}_6C$) isotope.

Thus, if one atom of $^{12}_{6}C$ has a mass of 12 units, it implies that 1/12 of it will have a mass of 1 unit.

NB.: We use the term one-twelfth because an atom of Carbon-12 is assigned an arbitrary value of 12.000a.m.u. Thus, it gives room for comparing elements that are lighter than the Carbon isotope. e.g. Hydrogen.

RELATIVE MOLECULAR MASSES

The masses of molecules of elements or compounds are also compared with the same standard as above. Thus, it is defined as follows:

The relative molecular mass of an element or a compound is the sum of the relative atomic masses of the atoms in one molecule of that compound.

For example, a molecule of Ammonium Chloride NH_4Cl has a relative molecular mass of 53.5 obtained as follows:

One atom of Nitrogen N, $A_r = 14 \times 1 = 14$

Four atoms of hydrogen H, $A_r = 1 \times 4 = 4$

One atom of Chlorine Cl, $A_r = 35.5 \times 1 = 35.5$

Thus, $NH_4Cl = 14 + 4 + 35.5 = 53.5$.

THE CONCEPT OF MOLE

Obviously, the relative atomic mass has no unit since it is the ratio of two numbers. However, if we introduce the unit of gram, g to the arbitrary value of Carbon-12 isotope which we said was 12.000amua.m.u, i.e 12.00g of $^{12}_{6}C$, we would be referring to a specific quantity of the element carbon. This specific quantity of course, would naturally consist of a specific number of Carbon atoms which is constant. This fixed number of atoms is now generally accepted after a series of investigations by that great scientist, **Avogadro**, as being **6.02 x 10²³** atoms and is called Avogadro's number or Avogadro's constant N_A. This specific number of atoms in 12.00g of Carbon-12 isotope has been adopted as a unit of the amount of substances and is known as a MOLE (symbol **n**).

Definition:

A mole is the amount of a substance which contains as many elementary entities as there are carbon atoms in 12.00 grams of carbon-12 isotope.

1 mole of $^{12}_{6}C$ = 12.00g of carbon-12 isotope and it contains 6.02 x 10^{23} atoms of $^{12}_{6}C$

In the same manner,

1 mole of substance **X**, weighing Y grams, also contains 6.02 x 10^{23} elementary entities of the substance **X**

Note that elementary entities refer to any type of particles which is a chemical constituent of the substance in question. These particles may be atoms, molecules, electrons, ions, neutrons, protons, etc. **A mole of any substance is therefore, the amount of that substance which contains 6.02 x 10^{23} particles of that substance.**

(Remember that 1 mole of electrons will carry 96,500 coulombs of charges).

From what we have discussed so far, you should be able to calculate the number of moles of any substance corresponding to a given mass of that substance and hence, the number of "elementary entities" it contains.

Worked Examples

1) A sample of lead has a mass of 18.63g.

i. How many moles of lead are present in the sample?

ii. How many atoms of the lead are present in the sample? (Given the relative atomic mass of lead to be 207).

$$No.\,of\,mole = \frac{Mass\,in\,grams}{Relative\,atomic\,mass\,of\,the\,element}$$

Answer:

i. Since the relative atomic mass of lead is 207, it implies that 207g of lead corresponds to one mole of lead.

Therefore, 18.63g of lead corresponds to:-

$\frac{18.63}{207}$ grams of lead = 0.09 moles of lead.

ii. Since one mole of any substance contains 6.02×10^{23} particles (in this case, atoms) it follows that 0.09 moles of lead contains $6.02 \times 10^{23} \times 0.09 = 5.4 \times 10^{22}$ atoms of lead.

2) What mass of Silver contains five times as many atoms as seven grams (7g) of Nitrogen? (Relative atomic masses are; Nitrogen 14 and Silver 108).

Answer:

Step i: Obtain the number of moles in 7g of Nitrogen

$$\frac{7}{14} = \text{moles} = 0.5 \text{ moles}$$

Step ii: Obtain the number of atoms of nitrogen in 0.5 moles of the element. $= 6.02 \times 10^{23} \times 0.5 = 3.01 \times 10^{23}$ atoms.

Step iii: Since the mass of silver required contains five (5) times the number of atoms as are in 7g of Nitrogen, then the number of atoms in such a mass of silver would be:

$$5 \times 3.01 \times 10^{23} = 1.505 \times 10^{24} \text{ atoms of silver.}$$

Step iv: Now convert the number of atoms of silver obtained in Step (iii) to number of moles by dividing that number of atoms by 6.02×10^{23}.

$$n = \frac{1.505 \times 10^{24}}{6.02 \times 10^{23}} = 2.5 \text{ moles.}$$

Step v: The mass of silver is obtained by multiplying the number of moles with the relative atomic mass of silver which is 108.

Therefore, the mass of silver required is:

$$2.5 \times 108 = 270g.$$

MOLAR VOLUME

The volume occupied by one mole of a gaseous element is known as the Molar volume of that element. The molar volume V_m is defined mathematically as:-

$$\frac{\text{The volume of a gas}}{\text{One mole of the gaseous element}} = \frac{V}{n} = V_m$$

The state of the gas must be specified. At S.T.P, the molar volume of any gas is **22.4 dm³**. That is, one mole of any gaseous element would at S.T.P, occupy a volume of **22.4 dm³**.

Practice questions for Students C2

1. In your own words, define atomic mass.

2. Define the relative molecular mass of a compound.

3. Calculate the relative molecular mass of the following molecules.

   (a) $(NH_4)_2 \, Ni \, (SO_4)_2 . (NH_4)_2 Ni(SO_4)_2$

   (b) $MgSO_4 . 7H_2O$

   (c) $KAg(CN)_2$

   (d) NaH_2PO_4

   (e) $AU_2(NO_3)_6$

4. What do you understand by ``a mole of a substance''? Define the term elementary entity.

5. Calculate the mass of gold that contains as many atoms as 13g of oxygen (relative atomic masses of gold and oxygen are 197 and 16 respectively).

CHAPTER THREE

THE CONCEPT OF OXIDATION NUMBER

INTRODUCTION

The atoms of all elements except noble gases have unstable electron arrangements. Unstable because their OUTERMOST electron shells contain less than (eight), 8 electrons and since all atoms want to attain stability, they will be willing to expel those electrons in excess of (eight), 8 in their outermost shell or accommodate electrons from other atoms to complete their octet arrangement. However, on many occasions, the atoms would neither expel nor accept electron(s) completely but would share some of their electrons with other atoms of the same element or of different elements. This phenomenon is referred to as chemical bonding.

When an electron is completely expelled or acquired by an atom, the net charge on that atom change; since under normal condition, (in neutral atoms,) the number of positive charges equals the number of negative charges. e.g.

$$Be + e^- \longrightarrow Be^-$$

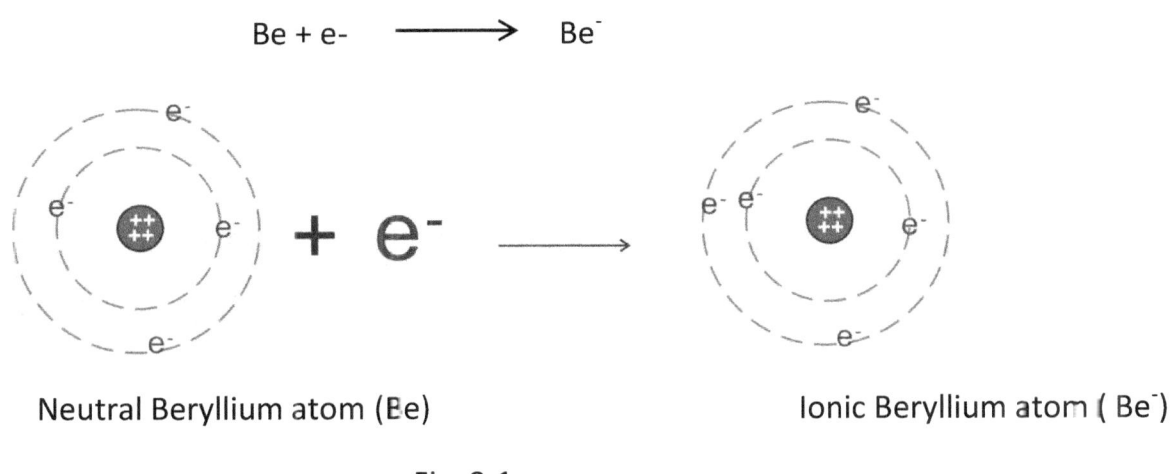

Neutral Beryllium atom (Be)   Ionic Beryllium atom (Be⁻)

Fig. 3.1

When two atoms share an electron, the electron would be shared equally by them if they are from the same element but would not be shared equally if they are from different elements; since one of the two atoms would be more electronegative(i.e.

electron loving) than the other one which is more electropositive. For example oxygen and hydrogen.

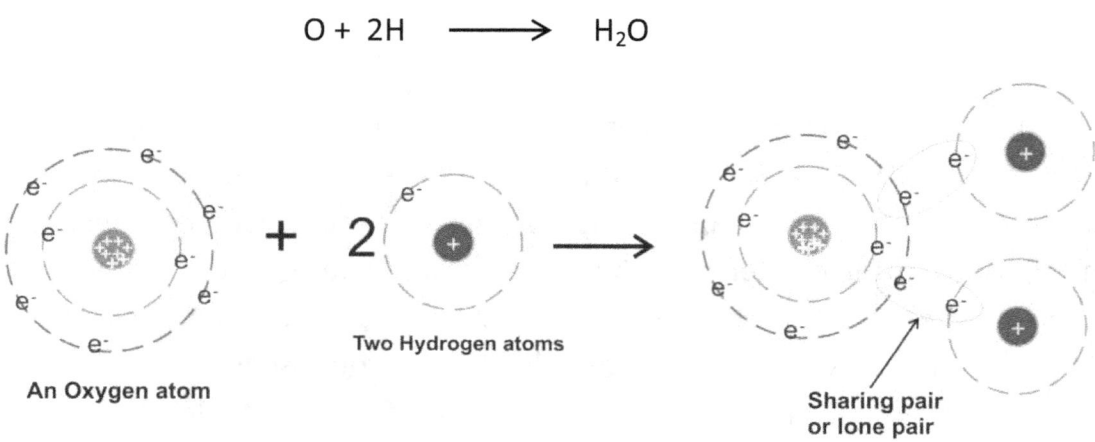

Fig. 3.2

Oxygen is more electronegative so, the electrons of the lone pair stay a longer time on oxygen than on hydrogen.

In such a case as above, the charge carried by that electron is temporarily assigned to the more electronegative atom in the molecule (i.e oxygen).
This assigned charge bestowed on the atoms what we call Oxidation number.
Let us now attempt to define oxidation number.

DEFINITION
Oxidation number is the number of CHARGE(S) which an atom in a molecule would have if the bonding electron(s) were assigned arbitrarily to the more electronegative element in the molecule.

EXPLANATION
A molecule is formed as a result of the combination of two or more atoms which may be from the same element (i.e. the same nature) or from different elements (i.e. different nature). This combination or bonding of atoms can take place in a variety of ways: two of which are prominent and relevant to our study here. These are:

IONIC AND COVALENT BONDING

The type of bonding found in a molecule is dictated by the nature of the atoms taking part in the combination.

IONIC BONDING

Ionic bonding is the result of a total transfer of electron(s) from one atom (the more electropositive of the atoms) to another atom (a more electronegative atom).

Generally, the group I and II element of the periodic table are more electropositive while the groups VI and VII elements are more electronegative. The combination of these categories will always produce an ionic compound.

For example, Na from group I and Cl from group VII

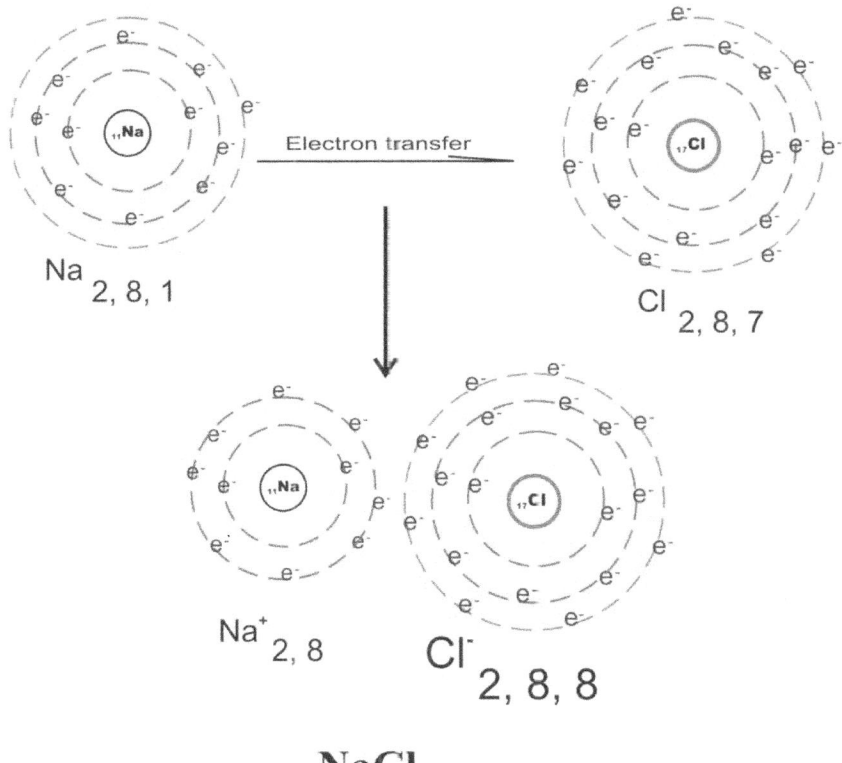

NaCl

Fig. 3.3

Ionic bond has been formed between Na and Cl.

Na + Cl ⟶ NaCl (ionic compound)

Ionic bond is the strongest of the "bondings" and it is not easily separated.

COVALENT BONDING

The covalent bonding on the other hand is the supposed equal sharing of pair of electron between the atoms concerned. Examples include Oxygen from group VI and Chlorine from group VII.

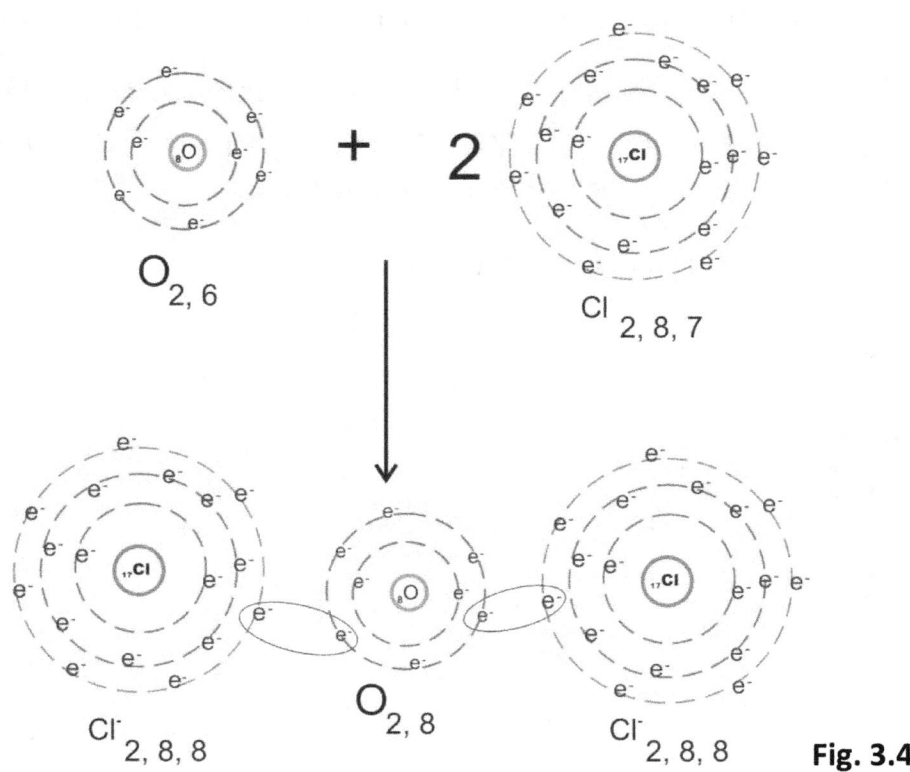

Fig. 3.4

Two covalent bonds have been formed between an atom of Oxygen and two atoms of Chlorine.

$$O_2 + 4Cl \longrightarrow 2Cl_2O \text{ (covalent compound)}$$

But, the sharing is only true when the atoms are of the same element. When the atoms involved are of different elements the electrons are always displaced towards the more electronegative of the two atoms concerned. Thus for the purpose of oxidation number, the displaced electron(s) is counted for the more electronegative atom.

OXIDATION NUMBER

The oxidation number represented with a charged number is indicated at the top right side of the atom concerned. When this is indicated on a compound, the charged number is the Arithmetic sum of the oxidation numbers of the individual atoms that makes up that formula (compound).

In assigning oxidation number, the following operational rules have to be followed:

1. In Free State, each atom of an element has an oxidation number zero (0). Thus for O_2, H_2, F_2, P_4 and S_8, $OX_n = 0$.

2. In ions which contain only one atom (i.e. monoatomic ions) the oxidation number of the element is equal to the charge on the ion.

Examples,

for Fe^{2+}, Fe^{3+}, K^+, Cl^-, S^{2-}, O^{2-} their oxidation numbers are respectively +2, +3, +1, -1, -2, -2.

It should be noted here that except for groups I and II elements which have +1 and +2 respectively, most other elements have variable oxidation numbers which depend on the compound in which they are found. E.g. Fe^{2+} in $FeCl_2$ and Fe^{3+} in Fe_2O_3.

3. The oxidation number of oxygen is -2 in most compounds where it is found except in the cases of peroxides where its oxidation number is -1. For example, H_2O_2 and Na_2O_2. Also in OF_2 where oxygen has an oxidation number of +2.

4. In compounds containing hydrogen, Hydrogen always have an Oxidation number of +1 e.g. H_2O, H_2SO_4, HF etc. except when it forms a metallic hydride when its oxidation number is -1. This is because hydrogen is more electronegative than most metals e.g. NaH, CaH_2.

5. In assigning oxidation numbers to atoms in a compound, the law of conservation of charges is applied. For, the sum of all the oxidation numbers of the atoms in a neutral molecule (i.e. an uncharged molecule) must be equal to zero. e.g. H_2SO_4

$H = +1$ $S = +6$ and $O = -2$

$H_2SO_4 = 2(+1) + 1(+6) + 4(-2) = 0$

However, in the cases of polyatomic ions (i.e. ions that contain more than one atom) the oxidation numbers of the atoms in such an ion adds up to the charge on the ion.

E.g. NO_3^-

$O = -2, N = +5$

$NO_3^- = 1(+5) + 3(-2) = -1$

This process can be used to assign oxidation numbers to atoms in a compound when their oxidation number is not certain.

E.g. $Pb(NO_3)_2$

$Pb = ?\ N = +5, O = -2$

$Pb(NO_3)_2 = ? + 2(+5) + 6(-2) = 0$

$Pb = -10 + 12 = +2$.

E.g. NH_4^+

$N = ?\ H = +1$

$NH_4^+ = N + 4(+1) = +1$

$N = +1 - 4 = -3$.

The above concept should not be confused with the concept of elemental valency. While oxidation number is the charge which an atom would have in a molecule if the bonding electron(s) were assigned arbitrarily to the more electronegative element in the molecule, a valence number is the combining power of an element which is a chemical property of the element in its own right.

In the Free State when an element has no oxidation number, it still has its valency. Thus, it can be said that oxidation number is not synonymous with valency even though group I element have valency of one (1), group II valency of two (2) and group VII have the valency of one (1). In the case of valency, all numbers do not carry charges; they are absolute.

Practice questions for Students C3

1. What do you understand by chemical bonding?

2. Define oxidation number.

3. State with suitable examples the five operational rules for assigning oxidation numbers.

4. We know that the oxidation number of oxygen is -2 in most of its compounds, why is its oxidation number +2 in OF_2 compound, -1 in Na_2O_2 and H_2O_2?

5. Hydrogen is known to have an oxidation number of +1 in most of its Compounds. When does it have -1 as oxidation number? Give three examples of such compounds.

6. What is the oxidation number of?

   (a) *Cr in Cr_2S_3 and CrO_3*

   (b) *K in $K_2Cr_2O_7$*

   (c) *Mn in MnO_4^{2-} and*

   (d) *S in $S_2O_8^{2-}$.*

CHAPTER FOUR

CHEMICAL NOMENCLATURE

In a community of people, individuals are identified by their names as well as the names of the functional groups they belong to.

In some communities, children are named on the eighth day of their birth. The name given to the child may reflect the sex, clan, and family occupation, day of birth, circumstance of birth and sometimes the future expectation of the child.

In Yoruba community of Nigeria for example, most names given to male children are not given to female children. So, you can distinguish males from females according to their names.

Names like Ayandele indicates a family trade of drumming, Abosede signifies that the child, (a female) was born on a Sunday, while a name like Asoroda signifies that this child cannot be betrayed but, if betrayed, there will be a consequence!

In the Hausa community, Dan Thanii indicates that the child was born on a Monday while Jumai indicates that the child, (a female) was born on a Friday. Also, a name like Dan Kano indicates that the child is rooted in Kano, etc.

These names were arrived at by following a set of rules already embedded in the culture and custom of such a community. So it is in the community of Chemistry too.

In 1921, a body known as International Union of Pure and Applied Chemistry (IUPAC) set up a commission of experts on the nomenclature of inorganic Chemistry. They were charged with the responsibility of formulating internationally acceptable names for inorganic compounds.

This commission submitted a set of rules and guidelines on the naming of inorganic compounds. These rules which have been revised several times are in consonance with the aims of this commission; and they are:

(1) To draft a set of rules that will guide the naming of inorganic substances.

(2) To ensure that the names arrived at by the applications of these rules must be clearly understandable to all chemists from all over the world.

(3) To ensure also that the names are simple and acceptable to all chemists irrespective of the language they speak.

These rules which we shall be meeting in the course of this book have been accepted worldwide. However, there are some difficulties in the understanding and application of the rules by beginners. Hence, the need for a few pages in this book to intimate them with their applications. We shall be concerned here with the naming of members of the following groups of chemicals:

(1) ions   (2) oxides

(3) acids   (4) bases

(5) salts (inorganic)   (6) halides.

[1] IONS

As you might have known, ion is the name given to charged substances These substances (ions) are classified into two types:

(A) The cations

(B) The anions.

(A) CATIONS: These are the ions that carry positive charges.

Mono atomic Cations

According to the IUPAC rules, mono atomic cations should be named like their corresponding neutral element with the word ion closing up the name.

e.g. *Na^+ (Sodium ion).*

In the case of elements with variable oxidation states, the appropriate oxidation number or state in Roman numeral should be included in brackets immediately before the word 'ion'.

e.g. *Fe^{3+} (iron (III) ion).*

Polyatomic cations:

Nitrogen based cations (i.e those that have their origin in Nitrogen).

(I) From Ammonia.

The cation formed by the addition of a proton to an ammonia molecule is called AMMONIUM ION and all substituted ammonium ions bear ammonium in their names. e.g.

$NH_3 + H^+ \longrightarrow NH_4^+$ Ammonium ion.

HON^+H_3 ———— Hydroxyl ammonium ion

$(CH_3)_3 - N^+H$ ———————— Trimethyl ammonium ion

(II) From Other Nitrogen bases.

Cations formed from other nitrogen bases like pyridine, hydrazine or aniline have the ending 'ium' added to the name of the base. e.g.

(C_5H_5N) ———————— Pyridine

$(C_5H_5N^+H)$ ———————— Pyridinium ion

H_2NNH_2 ———————— Hydrazine

$H_2NN^+H_3$ ———————— Hydrazinium ion

$C_6H_5 NH_2$ ———————— Aniline

$C_6H_5 N^+H_3$ ———————— Anilinium ion

In certain cases, the base can give more than one ion and as such, the appropriate charge is indicated in the name as follows:

e.g.

$N_2H_6^{2+}$ or $H_3N^+ - N^+H_3$ ———————— Hydrazinium (2+) ion.

N.B. Common names are used here. Since using their systematic (IUPAC) names will drag us into the organic nomenclature which is fully discussed in another volume of the series.

The Hydroxonium ion H_3O^+

Hydroxonium ion is a monohydrated proton. When a compound has H_3O^+ in it, or its derivatives, the name contains "oxonium"

E.g.

$H_3O^+ClO_4^-$ —————————— *Oxonium perchlorate*

$(CH_3)_2 OH^+$ —————————— *Dimethyl oxonium ion.*

Acidium ions

When a cation is formed by adding proton to an acid, the word ACIDIUM is included in the name of the corresponding acid, e.g.

$H_2NO_3^+$ —————————— *Nitrate acidium ion*

$H_2NO_2^+$ —————————— *Nitrite acidium ion*

CH_3COOH^+ —————————— *Acetate acidium ion*

(B) ANIONS

Anions are the atoms or groups of atoms that carry negative charge(s). Generally, the names of monoatomic anions end in 'ide'.

e.g. H^- —————————— *Hydride ion*

Cl^- —————————— *Chloride ion*

N^{3-} —————————— *Nitride ion*

Polyatomic anions

In the case of polyatomic anions, the name of the central atom is ended with 'ate' and its oxidation state is indicated in parenthesis before the word ion

e.g. NO_3^- Trioxo nitrate(V) ion.

However, there are some exceptions to this rule as there are some polyatomic anions whose names ends in 'ide' instead of 'ate'. A list of those anions is as below.

C_2^{2-} —————————— *Acetylide ion*

CN^-	——————	Cyanide ion
NH_2^-	——————	Amide ion
NH^{2-}	——————	Imide ion
$N_2^-\ N_3^-$	——————	Azide ion
$N_2H_3^-$	——————	Hydrazide ion
OH^-	——————	Hydroxide ion
$NHOH^-$	——————	Hydroxylamide ion
S_2^{2-}	——————	Disulphide ion
I_3^-	——————	Triiodide ion
O_3^-	——————	Ozonide ion
O_2^{2-}	——————	Peroxide ion
O_2^-	——————	Hyperoxide ion
Hf_2^-	——————	Hydrogen difluoride ion

There is a special class of anions whose names end in "O". This class of anions maintain their names even when in compound with some other elements.

e.g.
O^{2-}	——————	Oxo
OH^-	——————	Hydroxo
CN^-	——————	Cyano

When they are in compound with other elements their names are still prominent as in the following cases:

CO_3^{2-}	——————	Trioxo carbonate(IV) ion
$Cr_2O_7^{2-}$	——————	Hepta-oxo-dichromate(VI) ion
ClO^-	——————	Monoxo chlorate(I) ion
PO_4^{3-}	——————	Tetraoxo phosphate(V) ion

PO_4^{2-} ———————— Tetraoxo phospate(VI) ion

ClO_3^- ———————— Trioxo chlorate(V) ion

HSO_4^- ———————— Hydrogen tetraoxo Sulphate(VI) ion

NO_2^- ———————— Dioxo Nitrate(III) ion

SO_3^- ———————— Trioxo Sulphate(V) ion

Note that the above names are formed by first writing the name of the most electropositive member, followed by the number of oxo atoms using Greek prefixes and then the name of the central atom which upholds the rule of "ate" and the oxidation number of this central atom is indicated in brackets using Roman numeral.

[2] OXIDES

Oxides are a class of chemical compounds in which oxygen is in direct combination with another element. The element in combination with oxygen dictates the characteristics of the oxide. Thus, oxides can be divided into the following classes:

(a) Acidic oxides

(b) Basic oxides

(c) Amphoteric oxides

(d) Neutral Oxides

(e) Peroxides

(a) Acidic oxides

This is a class of oxides which exhibits acidic properties and they dissolve in water to form acid solution. These oxides are formed as a result of covalent bonding between oxygen and most elements found at the right hand side of the periodic table i.e. non-metallic elements. The molecules so formed are gases at room temperature and they are called acidic oxide or acid anhydrides (acid without hydrogen)

e.g. $S_{(s)} + O_{2(g)} \longrightarrow SO_{2(g)}$

$SO_{2(g)} + H_2O_{(l)} \longrightarrow HSO_3^- + H^+$

According to the IUPAC rule, their names end in "ide" and the oxidation state of the central atom is given in parenthesis immediately after the name of the central atom, then followed by the word oxide.

Note that ending the names of these oxides with "ide" is an exceptional condition to the rule of polyatomic nuclei e.g.

SO_2 — Sulphur (IV) oxide

NO_2 — Nitrogen (IV) oxide (formerly nitrogen dioxide)

N_2O_4 — Dinitrogen tetraoxide

CO — Carbon (II) oxide

(b) Basic Oxides

Basic oxides are the oxides formed through ionic bonding between oxygen and most elements to the left of the periodic table i.e. metals. Their names also end in "ide" and they dissolve in water to form bases.

e.g. $2Na + ½O_2 \longrightarrow Na_2O$

$Na_2O + H_2O \longrightarrow 2Na^+OH^-$

Examples of basic oxides are:-

Na_2O — Sodium oxide

CuO — Copper(II) oxide

Whenever the central atom of a compound has variable oxidation states, its state in the concerned compound must be indicated in parenthesis immediately after the name of the atom (as in CuO above).

(c) Amphoteric Oxides

These are oxides that cannot be sharply classified as either acidic or basic oxides. These oxides are those formed by the combination of oxygen and elements found around the center of the periodic table and they have intermediate behaviors which make them able to neutralize both acids as well as bases when they react with them.

e.g. **ZnO** ———————————— *Zinc oxide*

Al$_2$O$_3$ ———————————— *Aluminium(III) oxide*

As$_2$O$_3$ ———————————— *Arsenic(III) oxide*

Of course, their names end in **"ide"**.

(d) Neutral Oxides

This class of oxides results from combination of oxygen with some non-metals to form the appropriate non-metal oxide.

As the name indicates, they are neutral to both acids and bases since they do not neutralize either of them. Generally, they do not react with any other oxide. Examples include:

H$_2$O ———————————— *Hydrogen Oxide (Water)*

N$_2$O ———————————— *Dinitrogen Oxide*

CO ———————————— *Carbon (II)Oxide*

(e) Peroxides

These are oxides formed with bonds between the oxygen atoms in addition to bonds between the elements and oxygen. They react with dilute acids to form hydrogen peroxides. Examples include:

Ba$_2$O$_2$ ———————————— *Barium Peroxide*

H$_2$O$_2$ ———————————— *Hydrogen Peroxide*

Na$_2$O$_2$ ———————————— *Sodium Peroxide*

[3] ACIDS

An acid can be defined based on two schools of thought: Arrhenius (1884) said that an acid is a substance that can increase the concentration of hydrogen ion H^+ or hydroxonium ion H_3O^+ in aqueous solution.

Bronsted Lowry (1923) defined an acid as a substance that can donate a proton to some other substances. These two definitions are in fact saying the same thing but using different words.

e.g. $H_2SO_4 \longrightarrow H^+ + HSO_4^-$ **Bronsted Lowry**

$HCl + H_2O \longrightarrow H_3O^+ + Cl^-$ **Arrhenius**

From the above illustrations, we can see that an acid consist of two main parts i.e a cation (proton H^+) part and the anion part. According to IUPAC recommendation, for a two elements acid, the name should end in "ide" and the word acid be added.

e.g. *HCl* —————————————— *Hydrogen chloride acid*

HBr —————————————— *Hydrogen bromide acid*

In the case of acids with more than two elements, the name should end in "ate" like in the case of polyatomic anions (with more than one element). In fact, these acids are named exactly the way such anions were named except for the word "acid" which now replaces the word "ion".

e.g. *H_2PO_4* —————————————— *Tetraoxo phosphate (VI) acid*

H_2SO_3 —————————————— *Trioxosulphate (IV) acid*

Such acids are referred to as **Oxo acids**.

[4] BASES

A base may be defined as the metallic oxides or a metallic hydroxide which when in contact with an acid, neutralizes it to form salt and water.

From the definition above, we can see that bases can either be an oxide or a hydroxide of a metal.

Naming such compounds, we follow the same standard rules that have been discussed above when we were naming oxides; that is:-

For di-atomic nuclei, the name ends in "ide" starting with the name of the cation and indicating its oxidation state immediately after it (for those with variable oxidation states). e.g.

PbO —————————— Lead (II) Oxide

MnO_2 —————————— Manganese (IV) Oxide

ZnO —————————— Zinc Oxide

Similarly, metal hydroxides also known as soluble bases called Alkali are named as such.

$Zn(OH)_2$ —————————— *Zinc hydroxide*

NaOH —————————— *Sodium hydroxide*

KOH —————————— *Potassium hydroxide*

$Ca(OH)_2$ —————————— *Calcium hydroxide*

[5] SALTS (inorganic)

A salt is produced when the hydrogen of an acid is replaced by a cation (mostly metals). Simple (anhydrous) salts are the products of acid / base neutralization reactions.

Acid + Base $\longrightarrow$ Salt + Water

$HNO_3 + NH_4OH \longleftrightarrow NH_4NO_3 + H_2O$

The formation of salt can be seen from the above illustration as simply the substitution of the hydrogen ion of the acid by the cation of the base.

According to IUPAC recommendation, when naming a salt, the name of the cation is written first, followed by the number of oxo atoms present in the anion, then the name of the anion which ends in "ate" except when the salt is made up of two elements in which case the name of the anion should end in "ide". If the cation has variable oxidation states, its state in that compound in question is indicated in Roman numeral immediately after its name. Also, the oxidation state of the central element of the anion is indicated in Roman numeral in parenthesis after its name.

E.g. **NaCl** — *Sodium chloride*

Na_2CO_3 — *Sodium trioxo carbonate (IV)*

$CuSO_4$ — *Copper (II) tetraoxosulphate (VI)*

NH_4NO_3 — *Ammonium trioxonitrate (V)*

Fe_2SO_4 — *Iron (II) tetraoxosulphate (VI)*

$Fe_2(SO_4)_3$ — *Iron (III) tetraoxosulphate (VI)*

CLASSIFICATION OF SALTS.

For the purpose of this book, let us discuss the naming of four kinds of salt.

1. Acid Salts
2. Normal Salts
3. Basic Salts
4. Hydrated Salt

1. ACID SALTS

An acid salt is a salt in which not all the replaceable protons (hydrogen atoms) are displaced. When naming an acidic salt, the cation part is named first followed by hydrogen and then the anion part (starting with the number of oxo atoms) which should end in "ate". The oxidation state of the central element is indicated in the normal way.

e.g. **$NaHSO_4$** — *Sodium hydrogen tetraoxo sulphate(VI)*

$KHCO_3$ — *Potassium hydrogen trioxo carbonate(IV)*

2. NORMAL SALTS

A normal salt is that salt which does not contain any replaceable hydrogen atom i.e. all the replaceable hydrogen atom had been replaced by metal atoms. Such salts are named the same way as acid salts except that the word hydrogen will be omitted and the number of atoms of the cation present would be prefixed to the cation e.g.

Na_2SO_4 ——————— *Disodium tetraoxo sulphate VI)*

$MgSO_4$ ——————— *Magnesium tetraoxo sulphate (VI)*

3. BASIC SALTS

A basic salt on the other hand is a salt which contains undisplaced displaceable hydroxyl group. In naming basic salts, the cation part is named first as usual, followed by the other groups in alphabetical order. e.g.

$Mg(OH)Cl$ —— *Magnesium Chloride Hydroxide (because C comes before H)*

$CU_2(OH)_3Cl$ —— *Copper(II) Chloride trihydroxide (C comes before H)*

4. HYDRATED SALT

When naming salts with water of crystallization or with similarly loosely bounded molecules, the number of such molecules is generally designated by Arabic numeral before the name of such molecules. However, two other methods are equally acceptable. The salt portion is named like in the cases of anhydrous salts and it is followed by the description of the number of molecules of water (or similar molecule) present. e.g.

Method 1

$Na_2CO_3.10H_2O$ ——————— *Sodium trioxo carbonate (IV) decahydrate*

$FeSO_4.7H_2O$ ——————— *Iron (II) tetraoxo sulphate (VI) heptahydrate*

Method 2

$Na_2CO_3.10H_2O$ ——————— *Sodium trioxo carbonate (IV)-10- water*

$FeSO_4.7H_2O$ ——————— *Iron (II) tetraoxo sulphate (VI) -7-water*

Method 3

$Na_2CO_3.10H_2O$ —————— Sodium decaaquo trioxo carbonate (IV)

$FeSO_4.7H_2O$ —————— Iron(II) heptaaquo tetraoxo sulphate (VI)

[6] HALIDES

Halides are a class of chemical compounds which contain principally the Halogen group combining with other elements. Take the carbon halides for example (though this seems more of organic chemistry than inorganic, but a word about them will not do any harm to the objective of this book).

According to IUPAC rules they should be regarded as halomethane and the number of halogen atoms per molecule is specified by Greek prefix.

Cf_4 —————— Tetrafluoro methane

CCl_4 —————— Tetrachloro methane

$CHCl_3$ —————— Trichloro methane

CH_2Cl_2 —————— Dichloro methane

Further examples:

$CaCl_2$ —————— Calcium dichloride

$ZnCl_2$ —————— Zinc dichloride

$AlCl_3$ —————— Aluminum trichloride

A check list of IUPAC nomenclature of chemical formula is given in appendix III, arranged in an alphabetical order. More examples can be found there.

Practice questions for the Students. C4

1. State two of the rules for naming cationic compounds

2. State two of the rules for naming Anionic compounds.

3. Give the IUPAC names of the following compound.

   (a) $Ag(NH_3)_2^+$ (b) $Cu(NH_3)_4^{2+}$ (c) $S_2O_3^{2-}$ (d) MnO_4^{2-}

4. Oxygen is a very popular element; distinguish with illustrations, between its reactions with elements in the left hand side of the periodic table and those at the right hand side.

5. What class of oxide does titanium (IV) oxide belongs and what are the characteristics of this class?

6. Define an acid in your own words and state two rules for naming the various categories of it with examples.

7. (a) What is a salt?

   (b) Give four classes of salt and illustrate with examples.

8. Name the following compounds.

   (a) $Cu(NH_3)_4 Cl_2$

   (b) $CuSO_4 . 5H_2O$

   (c) HI

   (d) $K_2Cr_2O_7$

   (e) KClO

9. State five properties of soluble bases (Alkalis) and discuss two of their economic importance.

10. State in simple terms, how you will prepare an alkali in the laboratory.

CHAPTER FIVE

THE SPEAKABLE LANGUAGE

INTRODUCTION

The essence of language is to facilitate communication, promote understanding between the originator and recipient and enable useful, appropriate and efficient action to take place.

In many ways, Chemistry can be likened to a foreign language. To learn a new language, the student has to be taught the alphabets, words, sentences and grammar.

Incidentally, all these tools of a language are present in Chemistry.

Elemental symbols stand for alphabets.

Chemical formulae stand for words.

Balanced chemical equations stand for sentences and the rules for writing and balancing the chemical equations stand for grammar.

A chemical reaction equation is a symbolic summary of something that has happened or is proposed to happen in the laboratory. For example, when a solution of silver trioxo nitrate (v) and that of potassium chloride are mixed in the laboratory, we notice that some changes occur. The two clear solutions produce a white cloudy precipitate! This precipitate settles at the bottom of the test tube as a solid. When this solid is analyzed, it is found to be silver chloride AgCl. When the clear solution above the precipitate is also analyzed, it is found to contain potassium trioxo nitrate(v) KNO_3. This experiment which takes more than two days (fourty eight hours) to complete and more than half a page to report, can simply be summarized in one chemical sentence called chemical equation.

$$AgNO_3 + KCl \longrightarrow AgCl + KNO_3 \quad \text{_____} (1)$$

In this chemical shorthand (or sentence), the reagents we combine in the laboratory are written on the left hand side of the equation and they are called the REACTANTS while the substances that are produced from the experiment are written on the right hand side of the equation and are called the PRODUCTS.

CONSERVATION LAWS

Before a sentence is meaningful, the right punctuation has to be placed in the right place. So it is in chemical sentences too. An equation is meaningless if it is not balanced. A chemical equation is balanced only when it accords with the two laws of conservation i.e the law of conservation of matter and the law of conservation of charge. The law of conservation of matter may be stated thus:-

Atoms of an element can neither be created nor destroyed by chemical means.

The consequence of this law on balancing chemical equations is that every atom in the reactants must appear somewhere in the product and vice versa. Thus, equation (1) as above is balanced, since it obeys this law. That is a simple case anyway; since writing down the reactant and the products automatically balanced the equation.

However, for most reactions, the procedure for balancing the equations that summarize them would not be that simple; as multiples of some of the reactants and or the products may have to be used to make the number of each atom on both sides of the equation equal.

e.g.

$AgNO_3 + MgCl_2 \longrightarrow AgCl + Mg(NO_3)_2$ _____ (2a)

It can be seen that this equation is not balanced. Since we have two chlorine atoms on the left hand side and only one on the right hand side. But we can fix this by writing '2' as a coefficient of AgCl on the right hand side of the equation as below:

$AgNO_3 + MgCl_2 \longrightarrow 2AgCl + Mg(NO_3)_2$ _____ (2b)

A close inspection of equation (2b) shows that it is still unbalanced. This is because we now have two silver atoms on the right hand side and only one on the left hand side. This could be taken care of by writing a coefficient 2 in front of $AgNO_3$ on the left.

$2AgNO_3 + MgCl_2 \longrightarrow 2AgCl + Mg(NO_3)_2$ _____ (2c)

This is now a balanced equation which makes sense to anybody knowledgeable in Chemistry anywhere in the world.

REACTIONS IN SOLUTION (conservation law continues)

All soluble substances break up in their solutions to form ions. For example, many acids and salts dissolve in water to give ions. Salts like $AgNO_3$, KCl, KNO_3, and $Mg(NO_3)_2$ are all water soluble and are electrolytes. This means that in aqueous solution, they are dissolved to ions. AgCl however is an insoluble salt. If we write the ionic species present in the equation 2c above, we would have:

$$2Ag^+ + 2NO_3^- + Mg^+ + 2Cl^- \longrightarrow 2AgCl + Mg^+ + NO_3^- \qquad (3a)$$

In this equation, we see that Mg^+ and NO_3^- ion are present both as reactant and as products. Nothing happened to them during the reaction. Ions such as these; that do not participate in the reaction are called SPECTATOR IONS.

Hence, the net ionic equation of the reaction can be obtained simply by cancelling such ions out. Therefore we have:

$$Ag^+ + Cl^- \longrightarrow AgCl \qquad (3b)$$

This brings us to the second law of conservation, i.e. the law of conservation of the charges which states that:-

Charges cannot be created or destroyed in a chemical reaction.

The arithmetic sum of the charges on one side of the equation must be equal to the arithmetic sum of the charges on the other side. This law is obeyed in equation (3b) as:

$$(+1) + (-1) = 0$$

An important observation here is that the net ionic equation for both equations (1) and (2) above is the same. This leads us to conclude that any salt that dissociates to give chloride or any member of its family (halides) would react with silver nitrate $AgNO_3$ to give a similar chemical equation and hence, are the same reactions. We must, however, bear in mind that the conditions of similar looking reactions may be different from one another.

COMPLEX EQUATIONS

There are many reactions whose equations are so complicated that simple inspections procedure so far discussed would not be effective in balancing them. For these reactions, a more systematic procedure is employed.

Most complicated reactions belong to a particular class of reaction called OXIDATION REDUCTION REACTION or, simply REDOX reactions.

At this juncture, let us attempt to answer these important questions:

(1) What is a redox reaction?

A redox reaction is a chemical reaction in which the process of oxidation and the process of reduction take place simultaneously as the reaction progresses.

(2) What is oxidation and what is reduction?

Depending on the angle from which one wants to observe these processes, they may be defined in two ways:

(i) Oxidation is the process of electron releasing by an element or an ion while reduction is the process of electron gaining by an element or an ion.

(ii) Oxidation has taken place when there is an increase in the oxidation number of an element or an ion. While reduction has taken place when there is a decrease in the oxidation number of an element or an ion.

(3) What is a reducing agent and what is an oxidizing agent?

In terms of electron loss or gain, a substance releasing electron(s) is said to be a reducing agent (since it helps to reduce another substance) and itself is oxidized. On the other hand, the substance that accepts electron(s) is said to be an oxidizing agent (since it helps to oxidize the other substance by accepting its electron(s)) and itself is reduced.

From the above definitions, it is obvious that the subject of oxidation reduction reaction has to do with gain and loss of electron during the progress of the reaction. Thus, to investigate this class of reactions, we need to keep track of electrons in the chemical reactions.

There are basically two methods of balancing equations in this class.

(I) Half reaction or ion-electron method

(II) Oxidation number method.

Method I: Half reaction or ion-electron method.

Let us consider a reaction between metallic zinc and hydrogen chloride acid.

$$Zn_{(s)} + 2HCl_{(aq)} \longrightarrow ZnCl_{2\,(aq)} + H_{2\,(g)}$$

$$Zn + 2H^+ \longrightarrow H_2 + Zn^{2+} \qquad \text{_____ (4a)}$$

Matching up the appropriate atoms on either side, we see that neutral zinc atom (Zn) has changed into zinc ion Zn^{2+}

For this to occur, the Zinc atom must have lost two electrons. Zinc (Zn) has thus been oxidized.

$$Zn \longrightarrow Zn^{2+} + 2e^- \qquad \text{_____ oxidation process _____ (4b)}$$

On the other hand, each hydrogen ion on the left hand side of equation (4a) became part of a (neutral) hydrogen molecule. Here, each of the hydrogen ion must have gained an electron to become neutral. The hydrogen ion has thus been reduced.

$$2H^+ + 2e^- \longrightarrow H_2 \qquad \text{_____ reduction process _____ (4c)}$$

In this case, Hydrogen ion (H^+) is the oxidizing agent while Zinc (Zn) is the reducing agent. Remember that oxidizing agent is the substance that is reduced and the reducing agent is the substance that is oxidized.

Following the law of conservation of charges, in any chemical reaction, the electron lost must equal the electron gained. Thus oxidation and reduction must always be counterbalanced. This is the basis of the systematic method of balancing redox reactions.

What we simply do in this process is writing out the two half equation, maintaining an equal number of electron(s) on the two sides of each half equation and then adding the two half equations to get the net reaction equation.

e.g. $Zn \longrightarrow Zn^{2+} + 2e^-$ _____ Oxidation process

$\underline{2H^+ + 2e^- \longrightarrow H_2}$ _____ Reduction process

$Zn + 2H^+ \longrightarrow Zn^{2+} + H_2$ _____ (4d)

Sometimes when we have two half equations, the number of electrons lost in the oxidation process is not equal to the number of electrons gained in the reduction process.

e.g. $Al \longrightarrow Al^{3+} + 3e^-$ _____ oxidation process

$2H^+ + 2e^- \longrightarrow H_2$ _____ reduction process

In this type of situation, we multiply each half reaction by a factor that will equalize the number of electrons. In this case:

$2(Al \longrightarrow Al^{3+} + 3e^-) \quad = \quad 2Al \longrightarrow 2Al^{3+} + 6e^-$

and

$3(2H^+ + 2e^- \longrightarrow H_2) \quad = \quad 6H^+ + 6e^- \longrightarrow 3H_2$

Net reaction equation:

$2Al + 6H^+ + 6e^- \longrightarrow 2Al^{3+} + 3H_2 + 6e^-$ _____ (4e)

The systematic method for balancing redox reaction becomes most appreciated when the reactants and the products are more complicated. For instance, suppose that permanganate ion MnO_4^- reacts with iron (II) ion in acid medium to produce manganese ion and iron (III) ion.

We know that:

MnO_4^- produces Mn^{2+} _____ Reduction process

and Fe^{2+} produces Fe^{3+} _____ Oxidation process

$MnO_4^- + Fe^{2+} \longrightarrow Mn^{2+} + Fe^{3+}$

To balance these half reactions, we must balance not only the manganese and the charges, but also the oxygen. The systematic way of doing this is as follows and it must be memorized for the purpose of playing the game "HAT- the chemistry Teacher"

For All Reactions Taking Place in Acid Medium

(1) Balance the element gaining or losing electron(s) (in this case **Mn**)

(2) Balance the oxygen by adding water **(H_2O)** to the right-hand side.

(3) Balance the hydrogen introduced in step 2 by adding hydrogen ion H^+ to the left hand side.

(4) Balance the charge by adding electron(s) (to the receiving side)

In our example above, the steps would go this way:

Step 1: $MnO_4^- \longrightarrow Mn^{2+}$

Step 2: $MnO_4^- \longrightarrow Mn^{2+} + 4H_2O$

Step 3: $MnO_4^- + 8H^+ \longrightarrow Mn^{2+} + 4H_2O$

Step 4: $MnO_4^- + 8H^+ + 5e^- \longrightarrow Mn^{2+} + 4H_2O$

This is the reduction process _____ (5a)

The second part

$Fe^{2+} \longrightarrow Fe^{3+} + e^-$

has only one step since it is a simple oxidation. To balance electrons, multiply everything by 5. Hence,

$5Fe^{2+} \longrightarrow 5Fe^{3+} + 5e^-$ _____ (5b)

to obtain a net reaction equation, add equation (5a) to (5b) to get:

$MnO_4^- + 5Fe^{2+} + 8H^+ + 5e^- \longrightarrow Mn^{2+} + 5Fe^{3+} + 4H_2O + 5e^-$ ____ (5c)

For reactions taking place in basic medium

The process of balancing a redox reaction taking place in basic (alkaline) medium differ slightly from the above outlined steps; and it is a little bit more complicated than those taking place in acid medium.

In acid medium, the major species present are H^+ and H_2O; thus H^+ and H_2O are used to balance hydrogen and oxygen respectively. In basic medium however, the main species present are OH^- and H_2O. Since water contains more hydrogen than OH^-, we use H_2O to balance hydrogen and use OH^- to balance oxygen when a reaction occurs in basic medium. The following steps are taken when balancing a redox reaction taking place in basic medium:

(1) Balance the element gaining or losing e^-

(2) Balance oxygen by adding OH^- but add twice as many as needed.

(3) Balance hydrogen by adding H_2O to the left-hand side.

(4) Balance charges by adding electron(s).

The excess OH^- ion added in step 2 provides oxygen necessary for the water molecules to be added in **step 3.**

Note : *In a half equation, when electrons are placed at the left of the arrow, the substance is gaining electron and it is being reduced. While when electrons are placed to the right of the arrow the substance has lost electrons, hence, it is being oxidized.*

In some cases of reactions occurring in basic medium, a product or a reactant or both may contain hydrogen. In such a case, it is usually easier to balance the hydrogen first by adding H_2O and then balance oxygen by adding OH^-. Here, we add as many H_2O as twice the number of hydrogen needed.

e.g $\quad N_2H_4 \longrightarrow NH_3$

(1) Balance the element that is oxidized or reduced (in the above case, N oxidizes from N^{2+} to N^{3+}).

(2) Balance the hydrogen by adding H_2O add twice the number needed.

(3) Balance oxygen by adding OH^-.

(4) Balance the charges by adding electrons.

    Step 1: $N_2H_4 \longrightarrow 2NH_3$

    Step 2: $N_2H_4 + 2H_2O \longrightarrow 2NH_3$

    Step 3: $N_2H_4 + 2H_2O \longrightarrow 2NH_3 + 2OH^-$

    Step 4: $N_2H_4 + 2H_2O + 2e^- \longrightarrow 2NH_3 + 2OH^-$

METHOD II: OXIDATION NUMBER METHOD

The second method for keeping track of electrons in redox reaction is the oxidation number method (refer to the section on oxidation number). The oxidation number method can be used to balance redox equation by following these systematic procedures:

(1) Assign an oxidation number to each element in the reactants and the product except for those that clearly have the same oxidation number on both sides of the equation.

(2) Determine which element(s) change(s) oxidation number when reactants change to products, and calculate the change in oxidation number per unit of reactant used or product formed. (A decrease in oxidation number corresponds to reduction while an increase corresponds to oxidation.)

(3) Balance oxidation with reduction by using the appropriate multiplicative factors.

(4) Balance oxygen by adding H_2O (if reaction is in acid medium) or OH^- (if reaction is in basic medium).

(5) Balance hydrogen by adding H^+ for acid medium or H_2O for basic medium.

(6) Use the charge balance as a check, if the first five steps have been properly followed, the charges should balance.

Our example this time is the reaction of MnO_4^- and $H_2C_2O_4$ (oxalic acid). In acid medium, the products of these reactants are Mn^{2+} and CO_2. Thus, following the above outlined steps;

Step 1:

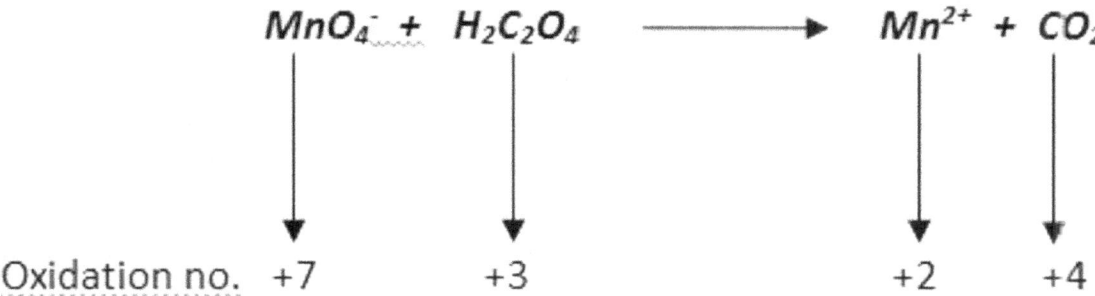

Step 2:

Mn changes from +7 to +2, a difference of -5 while carbon changes from +3 to +4, a difference of +1. But since we are dealing with charge per formula unit, we have to multiply the difference here by 2 (no. of atoms of C in one formula). This coefficient 2 is placed in front of the carbon (IV) oxide at the right-hand side of the equation to make up for those at the left-hand side.

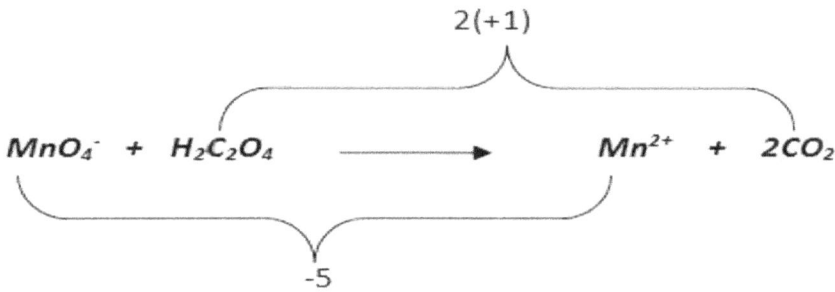

Step 3:
Multiply the pair MnO_4^- ⟶ Mn^{2+} by 2 and $H_2C_2O_4$ ⟶ $2CO_2$ by 5 to balance the reduction and oxidation.

$$2MnO_4^- + 5H_2C_2O_4 \longrightarrow 2Mn^{2+} + 10CO_2$$

Step 4:
Balance the oxygen by adding H_2O (since the reaction is taking place in acid medium)

$$2MnO_4^- + 5H_2C_2O_4 \longrightarrow 2Mn^{2+} + 10CO_2 + 8H_2O$$

Step 5:
Balance hydrogen by adding **6H⁺** to the left-hand side.

$$2MnO_4^- + 5H_2C_2O_4 \longrightarrow 2Mn^{2+} + 10CO_2 + 8H_2O$$

Step 6:

Check if the charges are balanced. If the above five steps were followed correctly, the charges would be balanced.

By now, I expect that you have mastered the process of equation balancing and I am confident that you would be able to handle any equation that comes your way. Below in Appendix (iii) is a check list of about one thousand balanced chemical equations. They have been grouped into various classes for easy access. However, some of them belong to many classes; but are grouped into a limited number of classes. Hence, this classification should not be seen as the end to it all.

Practice questions for Students. C5

1. State two conservation laws you know.

2. In your own words, what is a redox reaction?

3. Distinguish between an oxidation process and an oxidizing agent.

4. Use the methods learnt in this chapter to balance the following chemical reaction equations.

(i) $Mn_2O_{7(s)} \longrightarrow MnO_2 + O_{2(s)}$

(ii) $Ni(NO_3)_2 \longrightarrow NiO + NO_2 + O_2$

(iii) $Al_2O_{3(s)} + NaOH_{(aq)} + H_2O_{(l)} \longrightarrow NaAl(OH)_{4(aq)} + Al(OH)_3$

(iv) $Si_{(s)} + NaOH_{(aq)} + H_2O_{(l)} \longrightarrow Na_2SiO_{3(aq)} + H_{2(g)}$

(v) $TiCl_4 + Ti \longrightarrow TiCl_2$

(vi) $Fe^{2+} + H_2O + H^+ O_2 \longrightarrow Fe^{3+} H_2O + O_2$

(vii) $Cr_2O_7^{2-} + H^+ + Fe^{2+} \longrightarrow Cr^{3+} + Fe^{3+} + H_2O$

(viii) $FeCr_2O_4 + Na_2CO_3 + O_2 \longrightarrow NaCrO_4 + Fe_2O_3 + CO_2$

(ix) $MnO_4^{2-} + H_2O \longleftarrow MnO_4^- + MnO_2 + 4OH^-$

(x) $Au_2O_{3(s)} + NHO_{3(aq)} \longrightarrow Au_2(NO_3)_{6(aq)} + H_2O_{(l)}$

CHAPTER SIX

ELECTROLYSIS: ILLUSION OR REALITY?

INTRODUCTION

The discovery of Electricity marked a turning point in the history of our modern world. But more marveling is the effect of this electricity on chemical substances.

Experiments have shown that certain compounds decompose when electricity was passed through them. Why? And how?

This effect was first studied quantitatively by Michael Faraday in 1834. His findings formed the major base of our understanding of this phenomenon today.

Let us start from the first principle.

Under the study of Electricity, we classified all substances into two major groups: i.e those that allow the flow of electrons and thus conduct electricity, are called conductor while those that do not allow the flow of electrons and do not conduct electricity are called non-conductors or insulators. Another name for molten or liquid conductors is ELECTROLYTE while their non-conductor counterparts are called non-Electrolytes.

The chemical changes or processes which occur when electric current passes through an electrolyte are called Electrolysis. (Lysis - means decomposition).

It is safe therefore to say that an electrolyte is a substance which can conduct electric current and itself be decomposed by the current.

An electrolyte can conduct either in a fuse state or in solution and can therefore be classified as weak or strong electrolytes A weak electrolyte is a poor conductor, while a strong one is a good conductor of electricity.

In a set-up like Fig.6.1 below, which we use to demonstrate the process of electrolysis, the plates or wires that are dipped into the electrolyte and which connects it to the circuit (i.e. source of emf.) are called the ELECTRODES. The positive (+ve) electrode through which the current enters the solution (electrolyte) is called the cathode while other electrode i.e. the negative (-ve) through which current leaves the solution is called the anode.

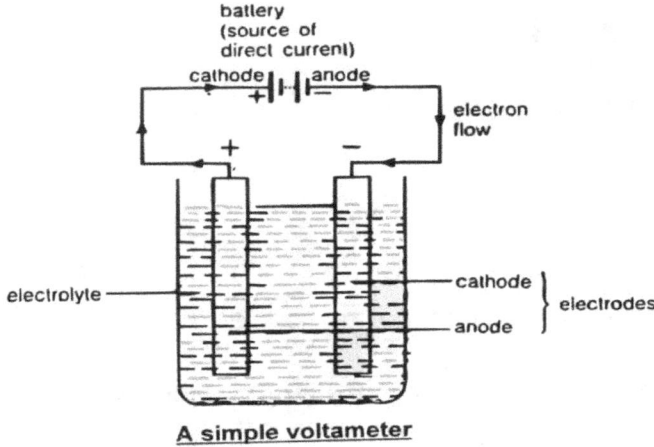

Fig 6.1

The whole arrangement is variously called electrolytic cell or voltameter.

If the electrolyte is a solution of a particular salt, base or acid, the voltameter may be called as such. E.g. for an electrolyte of a copper salt, it is called copper voltameter (do not confuse voltameter with voltmeter which is a voltage measuring instrument).

If it is acidulated water, it is called water voltameter - since it is the water that decomposes not the acid.

PROCESS OF ELECTROLYSIS

Basically, the conveyer of electricity in solids is electrons. However, in liquids, it is the ions. This is because, generally, solutions and molten solids are made up of ions moving about randomly but maintaining enough closeness with their partners and avoiding their likes in obedience to the Law - "Like repels; unlike attracts". Thus, when a potential difference is set up between the two electrodes (i.e. a source of e.m.f. is introduced) the positive (+ve) ions will move towards the cathode i.e. the negative electrode, while the negative ions will migrate towards the anode (the positive (+ve) electrode).

The migration of these two types of ions is as a result of the fact that the ions will want to go to where it can either take up electron(s) or where it will deposit its excess electron(s) in order to become neutral atom. However, since there are in most cases more than one kind of +ve ions or -ve ions in the solution, the question of which kind of the positive ions or -ve ions would be preferentially attended to at the electrode arises.

This question is solved by considering the respect or order exhibited by elements as shown in the electrochemical series below.

K	—	Potassium	F ----		Fluorine
Na	---	Sodium	Cl ----		Chlorine
Ca	—	Calcium	Br ----		Bromine
Mg	—	Magnesium	O -----		Oxygen
Al	—	Aluminum	I -----		Iodine
Zn	—	Zinc			
Fe	—	Iron			
Sn	—	Tin			
Pb	—	Lead			
H	—	Hydrogen			
Cu	—	Copper			
Hg	—	Mercury			
Ag	—	Silver			
Pt	—	Platinum			
Au	—	Gold			

INCREASING ACTIVITIES OF ELEMENTS (Non-metallic Elements)

INCREASING ACTIVITIES OF ELEMENTS (metallic Elements)

These series, which are further explained in the next chapter, shows that ions of elements below in the activity series will receive more attention at the electrodes than those above since the more active elements release their electrons more easily to form ions than the less active ones below them. For instance, a solution containing potassium, copper, iodine and chlorine ions will deposit copper at the cathode in preference to potassium and iodine will be liberated at the anode in preference to the chlorine. It should be noted that the result of the electrolysis of a solution of an electrolyte may be different from the result of electrolysis of the same electrolyte in a fused state.

It has been noticed that the nature of the electrode in use during electrolysis is important especially the anode. Whatever kind of cathode is used, metals will be deposited or hydrogen liberated at the cathode; but when an active anode is used, the result is that either an electronegative element is liberated at the anode in the order shown in the electrochemical series above or the electrode (of the anode) itself dissolves and goes into the solution since this requires less energy. For instance, if an electrode of copper metal is used in a solution containing copper ions, the copper anode will dissolve and go into the solution for subsequent migration to the cathode where it is deposited. However, when an inactive electrode like carbon is used as an anode, the electronegative elements are liberated at the anode following the order in the electrochemical series above.

Also, the concentration of the solution of electrolyte determines which element is liberated at the anode. A highly diluted solution of any electrolyte results in the liberation of oxygen at the anode in preference to any electronegative element but a highly concentrated solution results in the liberation of the other electronegative elements in accordance with the activity series. However, concentration does not affect the deposition of electropositive elements at the cathode.

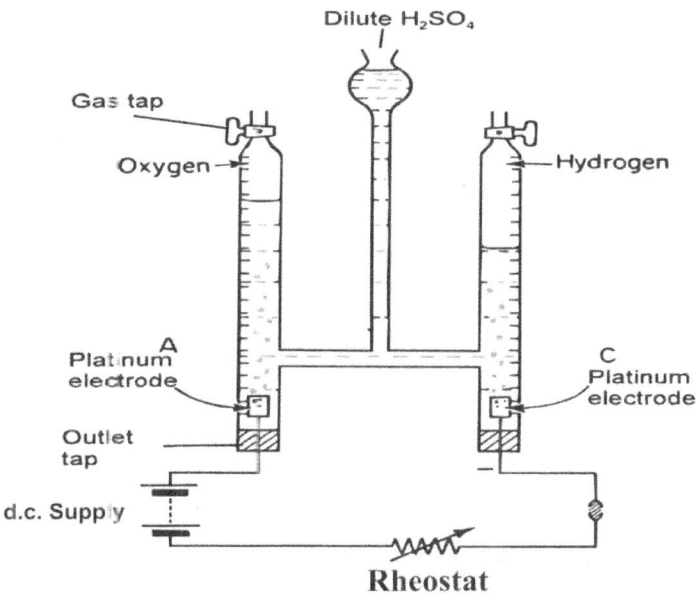

Hoffmann voltameter
(For electrolysis of water)
dilute acid is used.

Fig. 6.2

Laws of Electrolysis

Michael Faraday, in one of his numerous studies realized that when current was passed through a solution of copper tetraoxo sulphate VI ($CuSO_4$) using copper electrodes, copper was deposited on the cathode and lost from the anode.

He showed that the mass of copper dissolved from the anode by a given current of electricity in a given time is equal to the mass of copper deposited on the cathode in the same span of time. He also showed qualitatively that the mass dissolved or deposited is proportional to the product of the current I, passed through the solution and the time (t) for which it flowed. This implies that the mass increases as the product of current and time increase. Remember that I x t = Q where Q is the quantity of charge. So, the mass deposited is proportional to the quantity of charges that passes through the voltameter.

A similar result was obtained when he studied the electrolysis of water. He noticed that the variously liberated masses of hydrogen at the cathode and those of oxygen at the anode were each proportional to the quantity of charges that flowed in the voltameter. This led him to promulgate his FIRST law of electrolysis.

FIRST LAW:

The masses of any substance liberated or deposited during electrolysis is proportional to the quantity of electric charges that passed through the voltameter.

Mathematically:-

$$m \propto Q \quad \ldots\ldots\ldots (1)$$

$$m = ZQ \quad \ldots\ldots\ldots (2)$$

Where m represents mass, Q represent quantity of charge and Z is a constant called electrochemical equivalent (e.c.e) of the substance.

$$Z = \frac{m}{Q} \quad \ldots\ldots\ldots (3)$$

From the above equation (3), we can define electrochemical equivalent Z of a substance as the mass of the substance liberated or deposited during electrolysis per unit electric charge that passes.

But Q has earlier been defined as being equal to I x t,

or $Q = I \times t$ from above,

$$m = ZIt \quad \ldots\ldots\ldots (4)$$

and $$Z = \frac{m}{It} \quad \ldots\ldots\ldots (5)$$

Thus, the unit of **Z** is kg per coulomb or kgC^{-1}

Thus, if the electrochemical equivalent (e.c.e.) and the quantity of electric charges passed during the electrolysis are known, then the mass liberated or dissolved can be calculated using the above formulae.

SECOND LAW:

His second law which concerns the masses of different substances liberated by the same quantity of charges state that:-

The masses of different substances liberated or dissolved during electrolysis by the same quantity of electric charges are proportional to the ratio of their respective relative atomic masses to their electrovalency.

Let us put this definition in another form as follows:

The masses of different substances deposited or dissolved during electrolysis when the same quantity of electricity is passed through their different electrolytes are respectively proportional to their chemical equivalent. m_1 & Ce_1, m_2 & Ce_2

Note that the chemical equivalent of a substance is the quotient of its relative atomic mass to its electrovalency. i.e.

$$\text{Chemical equivalent} = \frac{\text{relative atomic mass of the element}}{\text{Electrovalency of the element}}$$

If the relative atomic mass is represented by **A**, and the electrovalency by **b**, then **Ce** (Chemical equivalent) is given as

$$Ce = \frac{A}{b}$$

The higher the value of a Ce, the higher the mass of the substance which the same quantity of electric charge will deposit.

Remember that electrovalency of an element is the no of charges (either **+ve** or **-ve**) carried by an ion of that element.

In honour of Michael Faraday, the quantity of electricity required to deposit, liberate or dissolve one mole of any substance is called FARADAY. This quantity is constant for all substances as seen in the second law of electrolysis above.

Recent calculation has shown that one Faraday (1F) is equal to 96,000 coulombs approximately and it is called Faraday constant symbol F.

Mechanism of Conduction in Electrolysis

The electrolytic conduction could be attributed to the splitting up of a compound into its ions in solution. The process of splitting is called ionization or ionic dissociation. The theory of conduction in electrolysis is credited to Arrhenius (1859-1927) even though Michael Faraday stated its essentials in 1834.

Arrhenius suggested that an electrolyte ionizes as soon as it was dissolved, that ions were not produced by current passed through the solution but it (the current) sets up a potential difference (p.d.) between the two electrodes which thus facilitates the migration of the already present ions in the solution to either of the electrodes.

For instance, a solution of silver trioxo nitrate(v) **$AgNO_3$** he supposed, contained silver ions **Ag^+** and trioxo nitrate(v) ions **NO_3^-**, and as such **Ag^+** ion migrates to the cathode (i.e. the negative electrode) where they pick up electrons and are deposited as silver atoms. At the same time, the **NO_3^-** ions migrates to the anode (i.e. the positive electrode) where they release their excess electrons.

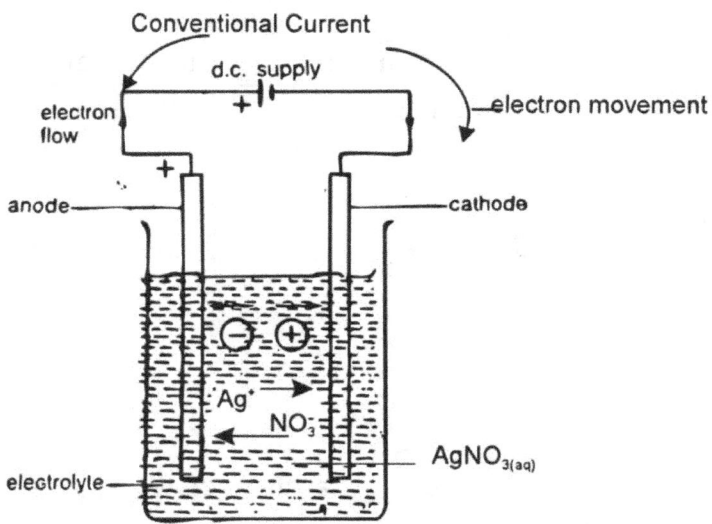

Fig. 6.3

When an active anode was used, i.e. metallic anode, a reaction like this;

$$NO_3^- + X^+ \longrightarrow XNO_3$$

Occurred and **XNO_3** molecules are formed where X is any metallic element usable as an electrolytic anode. [This is the method used to purify metals.] Thus, the anode get dissolved to release x^+ ions into the solution and the electron so released there from goes through the external circuit to the cathode where they are picked up by the arriving Ag^+ ions to be deposited as silver metal.

Where the anode is nonactive, i.e. passive anodes, oxygen gas is liberated while the NO_3^- ion combines with H^+ ions of the split water molecules H_2O to form HNO_3. Thus, the solution becomes acidic until the silver was completely deposited at the cathode. When this has been achieved, then, hydrogen would be liberated at the same cathode (in accordance with electrochemical series).

Salts of strong bases and acids such as silver trioxo nitrate (v) ($AgNO_3$), copper tetraoxo sulphate (vi) ($CuSO_4$), sodium chloride ($NaCl$) ionizes completely as soon as they are dissolved in water; i.e. the solution does not contain the molecules of these salts but only their ions. Such salts form strong electrolytes and so are the bases and acids from which they were produced since they also ionize completely when dissolved in water.

Some other salts like sodium trioxo carbonate (iv) Na_2SO_3 does not ionize completely when dissolved in water. They are salts produced from weak acids and are as such called weak electrolytes.

APPLICATIONS (USES) OF ELECTROLYSIS

Until the discovery of electrolysis processes, obtaining pure metals such as pure copper, pure silver or pure gold etc. was very cumbersome and are uncertain. However, now with the knowledge of electrolysis, the purest metals can be obtained with less effort.

As discussed earlier, the use of an active electrode of an impure metal as the anode and a thin strip of its pure metal as the cathode with a solution of the same metal's salt as the electrolyte is all that is needed.

Let us take copper purification for example. The system is set up as in fig.6.4.

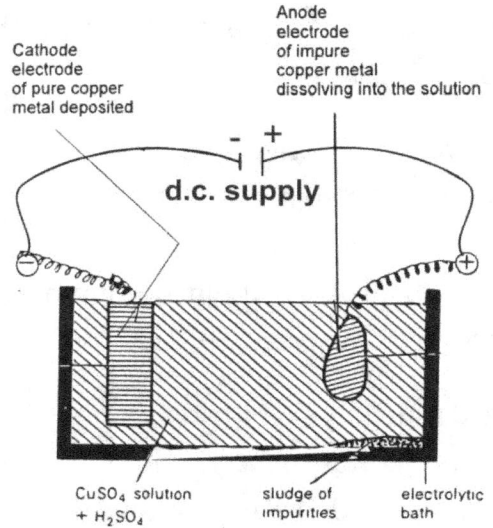

Electrolytic Bath

Fig. 6.4

The electrolyte used is copper II tetraoxo sulphate VI solution with a thin copper strip as the cathode and a bar of impure copper as the anode.

The ions present in the solution are:-

Cu^{2+}, SO_4^{2-} H^+ and OH^-

The mechanism is as follows.

Cu^{2+} and *H^+* migrates to the cathode where *Cu^{2+}* is preferentially discharged and thus deposited on the cathode electrode as pure copper metal.

At the anode, electrons are extracted from the atoms of the impure copper and channeled through the circuit while the copper ions *Cu^{2+}* resulting from this are passed into the solution from where they migrate to the cathode as earlier discussed.

The concentration of the solution thus remains constant until the whole of the anode is dissolved. The impurities present in the impure copper settles at the bottom of the voltaic cell as sludge.

The same process above can be used to effect the plating or coating of some articles of inferior metal with some highly valued metals such as silver or gold. This process is called **ELECTROPLATING** and it can be defined as **the process of depositing or coating an article with a thin layer of metal, using the processes of electrolysis.**

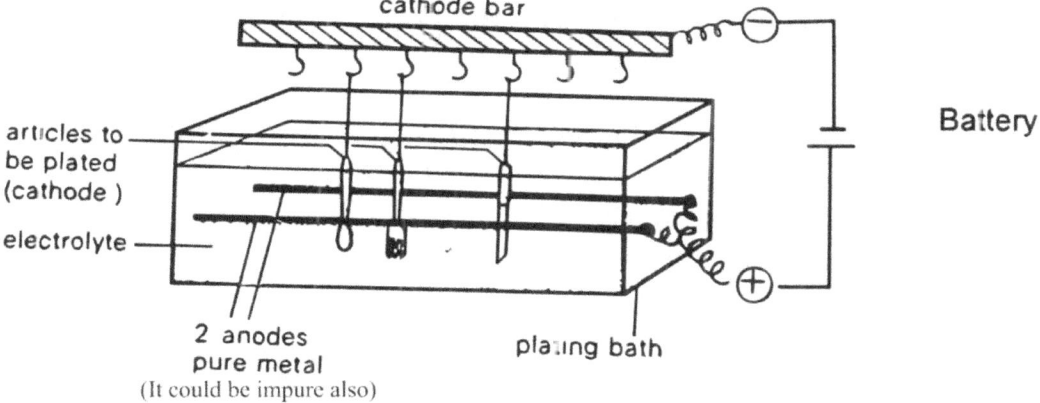

An Electroplating set up

Fig. 6.5

The articles to be electroplated are first plated with copper using copper plating bath, in which the anode is pure copper and the electrolyte is acidified copper II tetraoxo sulphate VI Solution; after which the copper plated articles are then made the cathode of a NICKEL-plating bath with a rod of pure nickel as the anode and a solution of Nickel II ammonium tetraoxo sulphate (VI) $(NH_4)_2 Ni(SO_4)_2$ for an article to be plated with nickel.

For articles to be plated with silver, a silver plating bath with a rod of pure silver as the anode and an electrolyte of silver trioxo nitrate (V) $AgNO_3$ or an electrolyte of potassium dicyano silver $KAg(CN)_2$ solution is used.

For articles to be plated with chromium, a chromium plating bath with pure chromium as the anode and an electrolyte of chromic acid like chromium II tetraoxo sulphate VI is used.

For articles to be plated with gold, a gold plating bath with a rod of gold as the anode and an electrolyte of gold III trioxo nitrate (V) $AU(NO_3)_3$ is used.

Effects or Applications of Electroplating
1. It increases the attractiveness of the articles (e.g. gold and silver plating).

2. It helps to prevent corrosion of the article (e.g. silver and chromium plating).

3. It gives a hard surface to the article (e.g. chromium plating).

4. It gives thickness to worn parts of machinery (e.g. nickel plating).

Electrolysis process is very useful in the commercial preparation of metals and non-metals. e.g. metals like Potassium (**K**), Sodium (**Na**) Calcium (**Ca**) are obtained by electrolysis of their fused salts.

The electrolysis of brine (common salt solution) produces chlorine in a special set up called mercury cathode cell.

There are so many other applications of electrolysis, such as in - electrolytic capacitors, the rechargeable batteries (accumulators) etc. which are in the jurisdiction of the physicists.

Practice questions for the Students C6

1. Why is it that either aqueous solution or fused **NaCl;** conducts electricity but solid **NaCl** does not?

2. State Faraday's laws of electrolysis?

3. Give a detailed account of what happens at the anode and cathode when active electrodes are used.

4. In not more than six sentences, describe the process of electroplating using suitable examples.

5. During a process of electrolysis, it was discovered that 2.11g of Copper was deposited in 52 mins. What quantity of current was involved?

(Take electrochemical equivalent of copper to be 0.0033g/c and its chemical equivalent is 31.8)

CHAPTER SEVEN
DEFINITIONS OF SOME REACTION PROCESSES

Laws, Concepts and Principles.

(1) A COMPOUND
A compound is a substance which contains two or more elements chemically combined together. A compound is formed as a result of a chemical change and it is a new substance with entirely different properties from those of its component elements or substances from which it was formed.

(2) DECOMPOSITION REACTION
This is a fundamental type of chemical change in which one substance breaks down into two or more simpler substances.

e.g. $2H_2O \longrightarrow 2H_2 + O_2$

or in the case of double decompositions, two compounds break down and recombine to form two or more different compounds.

e.g. $2HCl + CaCO_3 \longrightarrow CaCl_2 + H_2CO_3$

(3) DEHYDROGENATION REACTION
This is a process whereby hydrogen is removed from a compound by chemical means.

(4) DISPLACEMENT REACTION
This is a process of chemical reaction where metal ions displace one another from their compounds in solution in accordance with the rule of electrochemical series.

ELECTROCHEMICAL SERIES
In their effort to investigate the ease with which elements lose electrons, chemists carried out a great number of experiments which led to the establishment of what is now known as the activity series. In this activity series, metallic elements are arranged in an order that each element in the series loses - electrons more easily than the element below it and as a result, exhibits greater chemical activity as a metal by displacing from solution those elements below it and itself is displaced by those above it.

There are two series, the metallic elements and the non-metallic elements; and they are as follows:

Activity Series (Metallic elements)

	Metal	Symbol
1.	Potassium	K
2.	Sodium	Na
3.	Calcium	Ca
4.	Magnesium	Mg
5.	Aluminum	Al
6.	Zinc	Zn
7.	Iron	Fe
8.	Tin	Sn
9.	Lead	Pb
10.	Hydrogen	H
11.	Copper	Cu
12.	Mercury	Hg
13.	Silver	Ag
14.	Platinum	Pt
15.	Gold	Au

Activity Series (Non-metallic Elements)

	Element	Symbol
1.	Fluorine	F
2.	Chlorine	Cl
3.	Bromine	Br
4.	Oxygen	O
5.	Iodine	I

Hydrogen, though nonmetallic, is placed in the metallic activity series to indicate its relationship with the metals. Since the metals above it in the series will liberate it from acids while those below it do not.

(5) DISPROPORTIONATION REACTION

A chemical reaction in which a single compound serves as both oxidizing and reducing agents; and it is thereby converted into a more oxidized and a more reduced derivative.

(6) ELECTRON

Electron is a fundamental particle of matter that can exist either as a constituent of an atom or in the Free State. It has a negative electric charge. Electrons can be removed from the atoms of metals and some other elements by heat, light, electric energy, or bombardment with high-energy particles (Radiation, ionization). In such cases, they are totally free from the atomic orbit and their energy can be utilized by means of a conductor (electricity) or a vacuum tube or a semiconductor (in electronics).

(7) ELEMENTS

An element is a substance that comprises of only one type of matter. i.e. they cannot be split into simpler substances or simpler units by ordinary chemical process. There are about 105 such substances known to man presently. The smallest unit of any element is the atom. All the atoms of a given element are identical in nuclear charge and number of electrons and protons, but they may differ in mass (isotopes). All elements beyond lead are unstable and are radioactive. Note that all elements beyond uranium (i.e transuranic) were nonexistent in 1940. They were artificially created by neutron bombardment of other elements (cyclotron).

(8) ELECTRO NEGATIVITY

All atoms that have fewer than eight (8) electrons in their highest (or outermost shell) principal quantum level (except for helium which is a noble gas) have low energy orbital vacancies capable of accommodating electrons from outside the atom. The existence of these vacancies is evidence that within these regions, the nuclear charge can exert a significant attraction for such electron(s) even though as a whole, the atom is electrically neutral. This attraction is called ELECTRO NEGATIVITY.

(Electro negativity may then be said to be the tendency to become negatively charged.)

(9) HYDROGENATION REACTION

This is the chemical combination of hydrogen with another substance, usually an unsaturated organic compound.

(10) HYDROLYSIS (HYDROXIDATION)

This is a chemical reaction in which water reacts with another substance to form one or more new substances. This involves splitting of the water molecule into ions. For example, reaction of ions of dissolved salts to form various products such as acids, complex ions etc.

(11) ION EXCHANGE

This is a reversible chemical reaction between a solid (ion exchanger) and a fluid (usually an aqueous solution) by means of which ions may be interchanged from one substance to another.

(12) IONIZATION

This is the separation or disassociation of a molecule into ions of opposite electrical signs. This occurs spontaneously in many salts when dissolved in water or melted.

(13) NEUTRALIZATION REACTIONS

These are the reactions between hydrogen ion from an acid and hydroxyl ion from a base to produce salt and water or in non-aqueous solvents, the reactions between the positive and negative ions of the solvent to produce solvent and another salt-like compound.

(14) OXYGENATION REACTIONS

This is a reaction in which oxygen combine chemically with another substance to form an oxide. This process may sometimes be referred to as combustion in oxygen.

(15) REVERSIBLE REACTION

These are reactions in which the reactants of a reaction are obtained as the product of the same reaction when the conditions of the forward reaction are reversed and the products of the forward reaction become the reactant(s). They are usually represented by two half headed arrows facing opposite directions.

(16) SUBSTANCE

A substance in chemistry is any chemical element or compound. All substances are characterized by a unique constitution and are thus homogeneous i.e. a material of which every part is like every other part is said to be a substance.

(17) SUBSTITUTION REACTIONS

This is the replacement of one element or radical by another as a result of a chemical reaction.

(1) THE MOLE

The mole is one of the seven fundamental units in the International System of Units **(SI)**. The mole is the unit used for measuring the amount of a substance and is defined as the amount of a substance containing the same number of atoms, molecules, or ions as the number of atoms in **12 grams** of ^{12}C. Because there are **6.022 × 10^{23}** atoms of carbon in **12 grams** of ^{12}C, this number **(6.022 × 10^{23})**, known as Avogadro's number, is the amount of matter containing **6.022 × 10^{23}** atoms, molecules, or ions.

The mole concept provides a means of calculating how many atoms, ions, or molecules are in a sample by weighing the substance. From the definition of atomic weight, the amount of any element that has a mass (in grams) equal to its atomic weight (available on the periodic table) will contain **6.022 × 10^{23}** atoms. Thus, **4.0026** grams of helium, **32.0064** grams of sulfur and **200.59** grams of mercury each contain **6.022 × 10^{23}** atoms.

Similarly, a mole of a molecular substance **(6.022 × 10^{23} molecules)** is the amount of the substance whose mass (in grams) is equal to its molecular weight. Molecular weight is derived by summing the atomic weights of the atoms composing a molecule. For example, **70.906 grams (2 × 35.453)** of Cl_2 contains **6.022 × 10^{23}** molecules (one mole) of ^{12}C.

Chemists use the same principle to measure the number of ions in a compound. For example, one mole of sodium ions **(Na^+)** has a mass of **22.9898** grams (atomic weight of **Na is 22.9898**). One mole of **NaCl** has a mass of **58.443** grams **(22.9898 + 35.453)**.

(2). MOLARITY

Molarity is the concentration of a substance in solution and is expressed as moles of solute per liter of the solution. Thus, a **0.1** molar (abbreviated **0.1 M**) solution of sodium chloride contains **5.8443 (58.443 × 0.1)** grams of **NaCl** per liter of solution.

(3). MOLALITY

Molality, a term less frequently used than molarity, is the number of moles of solute in 1000 grams of solvent. Thus, a **0.1** molal solution of sodium chloride in water has **5.8443** grams of **NaCl** in **1000** grams of H_2O.

(4). NORMALITY

Normality is the number of equivalents per liter of solution. For acid-base-salt systems, an equivalent is the amount of the substance that will gain or lose one mole of H^+ ions. For instance, one mole of sulfuric acid **(H_2SO_4)**, which has a mass of **98.0795** grams, produces two moles of **H⁺**, or two equivalents. Therefore, a one molar solution of sulfuric acid is a two normal **(2 N)** solution. A **0.1 N** solution (containing **0.1** moles of **H⁺**) of sulfuric acid contains **4.90397** grams of H_2SO_4 per liter of solution **([98.0795/2] × 0.1)**.

LAWS

1. Law of Conservation of Mass [Lavoisier 1794]

Lavoisier, a French chemist observed in 1794 that the total mass of all products of a chemical reaction was the same as the total mass of the reacting substances. This observation led him to this law.

Matter is neither created nor destroyed in the course of a chemical reaction.

2. Law of Constant Composition or Definite Proportion [Proust 1779]

Proust, a French chemist, carried out a series of experiment from which he observed that when reactions take place between two substances **A** and **B**, and **1cm³** of **A** reacts completely with **2cm³** of **B**. Then, **2cm³** of **A** will also react completely with **4cm³** of **B**.

If however, **2cm³** of **A** were added (reactant) to **5cm³** of **B**, only **4cm³** of **B** will participate in the reaction leaving **1cm³** of **B** unreacted. This led him to promulgate his law which states that:-

All pure samples of the same chemical compound contain the same elements combined together in the same proportion by mass.

3. Law of Multiple Proportions [John Dalton 1803]

Dalton, an English chemist, in 1803 investigated the composition of different compounds formed from the same elements. Compounds like copper oxides formed only from copper and oxygen for example has two compounds i.e. copper II oxide and copper I oxide. On reduction of this compound (oxides) using hydrogen, the masses of copper formed were measured and it was observed that for equal masses of the respective oxides, the masses of the copper obtained were in a simple ratio.

In one of his experiments, he observed that 10g of oxygen combined with 40g of copper to form copper II oxide and 10g of oxygen combined with 80g of copper to form copper I oxide, which shows that the masses of copper which combined with 10g of oxygen were respectively 40g in **CuO** and 80g in **CuO₂** i.e. in ratio 1:2. These observations led him to promulgate this law which states:-

When two elements A and B combine together (in more than one proportion) to form more than one chemical compound, then the different masses of A which separately combine with a fixed mass of B are in the ratio of small whole numbers.

4. The Law of Equivalent

This law which has its origin from the law of reciprocal proportion (Richter 1792) and the Law of multiple proportions (Dalton 1803) state:-Elements combine together or replaces one another in the ratio of their equivalent masses.

But, **the equivalent mass of an element is defined as the number of parts by mass of an element that combines with or replaces 8 parts by mass of oxygen.**

That is, on O = 8 scale, the equivalent of hydrogen is 1.0080g and that of chlorine is 35.457g since the mass of chlorine will combine with 1.0080g of hydrogen.

Thus, the equivalent mass of an element = $\dfrac{\text{its atomic mass}}{\text{its valency}}$

5. Ionic Theory

It State that: Electrolytes split into electrically charged ions when in solution or fused state.

6a. Gay-Lussac's Law of Combine Volume [1808]

It states that: When gases react, they do so in volumes which bear a simple ratio to one another and to the volume of any gaseous product(s) formed if all measurements are under the same conditions of temperature and pressure.

6b. Avogadro's Hypothesis [1811]

It states that: Equal volumes of all gases if measured under the same conditions of temperature and pressure, contains the same number of molecules.

7. Since Avogadro's law, more or less explains Gay-Lussac's law of combine volume, it may be better to treat the two together as follows:

If the temperature and pressure are held constant, then a volume **5.00cm³** of hydrogen will contain the same number of molecules as **5.00cm³** of chlorine, (Avogadro's), thus, when they combine, they do so in volume (Gay lussac's)

1 vol. of Hydrogen + 1 vol. of Chlorine ⟶ 2 vol of Hydrogen Chloride gas.

This implies from Avogadro's hypothesis that:

n molecules of hydrogen + n molecules of chlorine ⟶ 2n molecules of hydrogen chloride gas.

Therefore if **n = 1**, then we have;

One molecule of hydrogen (H₂) + one molecule of chlorine (Cl₂) ⟶ two molecules of hydrogen chloride 2HCl.

Symbolically; $H_2 + Cl_2 \longrightarrow 2HCl$

8. Boyles' Law

An elementary understanding of the Kinetic theory of matter shows that molecules are in continuous random motion in any container. They bombard the walls of the containing vessel continuously and as such bring pressure to bear on the system.

Since $\text{pressure} = \dfrac{Force}{Area}$

Force = rate of change of momentum in a given direction, thus, if the volume of the gas is compressed from V_1 to V_2 at constant temperature, the number of molecules occupying that volume will then be more than it was. If their kinetic energies remain the same, they will travel less distances per unit time to bombard the walls of the vessel and many more collision per unit time will occur. Hence, the rate of change of momentum of the molecules will increase leading to an increase in the pressure generated.

Mathematically,

Rate of colliding molecules per unit area is inversely proportional to the volume.

R.C. M .P.U.A   α   1/V

or force per unit area α 1/V

∴ Pressure α 1/V

This was the bases of Robert Boyle's law which states that: *at constant temperature, the volume of a gas is inversely proportional to its pressure.*

i.e the volume decreases as the pressure increases or pressure increases as volume decreases.

So   P α 1/V

&   P = K 1/V

∴   PV = K (where k is a constant)

So that $P_1V_1 = P_2V_2 = P_3V_3$ etc.

9.   Charles' Law

In a similar manner, a French scientist, Jacques Charles came up with his popular law named after him which states that:

At constant pressure, the volume of a given mass of gas is directly proportional to its absolute temperature (thermodynamic)

Or more specifically stated;

At constant pressure, the volume of a given mass of gas expands by *1/273* of its volume at *0°C* per kelvin rise in temperature. That means that the volume increases as the absolute temperature increases.

Thus: *V α T*

or   *V = KT*

and   *V/T = K (Constant)*

so that $V_1/T_1 = V_2/T_2 = V_3/T_3$ *etc.*

10. Pressure Law

Similarly, a study of the effect of change in the absolute temperature of a fixed mass of gas on its pressure at constant volume shows that the pressure of a fixed mass of gas at constant volume is directly proportional to the absolute temperature.

Thus,

$P \alpha T$

or $\quad P = KT$

and $\quad P/T = K$ (Constant)

So that $\quad P_1/T_1 = P_2/T_2 = P_3/T_3$ etc.

(10b) The above laws, i.e Boyle's, Charles' and Pressure Laws can be combined to obtain the general gas law as follows:

If T is constant, then PV = constant k.

If P is constant, then V/T = Constant K

If V is Constant, then P/T = constant K.

$$\therefore \frac{PV}{T} = \text{Constant K.}$$

Thus; $P_1V_1/T_1 = P_2V_2/T_2 = P_3V_3/T_3$ etc.

11. Dalton's Law of Partial Pressure [1810]

Dalton (1808), an English Chemist, noticed that when two or more gases are mixed together, the total pressure exerted by the mixture is equal to the sum of the pressures which each separate gas would exert if it occupies the space alone. The pressure of each of the gases is called partial pressure.

We may simply define partial pressure of a gas in a mixture as the pressure which that gas would exert if it alone occupied the same volume.

Each gas in a mixture of gases has the same pressure as it would exert if it were present alone in the volume occupied by the mixture.

The law simply state that the pressure exerted by a mixture of gases is equal to the sum of the partial pressures of its component gases.

12. Graham's Law of Diffusion

Graham, a Scottish chemist, 1846 found that if a gas was allowed to escape from its container through a small hole into a vacuum, the rate at which molecules of the gas escape will depend upon the rate at which they reach the area represented by the hole. This process of escape of gas molecules into a vacuum through a small hole is called **EFFUSION.**

Diffusion on the other hand is the passage of a gas through a porous partition i.e. with a large number of tiny holes through which the molecules can escape instead of a single one.

Graham's law does states that; the rates of diffusion or (effusion) of gases under the same given conditions are inversely proportional to the square roots of the r densities.

This means that if two gases are allowed in turn to pass through the same effusion hole under identical temperature and pressure conditions and the time taken for the same volume of each to pass through are compared, it would be found that,

$$\frac{Time\ taken\ for\ gas\ A}{Time\ taken\ for\ gas\ B} = \frac{1}{\sqrt{\frac{density\ of\ gas\ A}{density\ of\ gas\ B}}}$$

$$\frac{Rate\ of\ diffusion\ of\ gas\ A}{Rate\ of\ diffusion\ of\ gas\ B} = \sqrt{\frac{Density\ of\ gas\ B}{Density\ of\ gas\ A}}$$

Or;

$$\frac{Rate\ A}{Rate\ B} = \sqrt{\frac{d_B}{d_A}} \quad \text{and} \quad \frac{Rate\ B}{Rate\ A} = \sqrt{\frac{d_A}{d_B}}$$

But density is mass per unit volume i.e. **d = M/V** substituting for densities,

$$\frac{Rate\ B}{Rate\ A} = \sqrt{\frac{\frac{M_A}{V_A}}{\frac{M_B}{V_B}}} = \sqrt{\frac{M_A}{V_A} \times \frac{V_B}{M_B}}$$

Since the volumes of the gasses were the same, i.e $V_A = V_B$

Hence,

$$\frac{Rate\ B}{Rate\ A} = \sqrt{\frac{M_A}{M_B}}$$

13. Le Chatelier's Principle [1885]

When constraint is applied to a system in equilibrium, the system will change (adjust itself) in such a way as to counteract (or remove) the effect of such constraint.

There are three major constraints that systems in equilibrium may experience. These are:

(a) Change in temperature

(b) Change in component concentration

(c) Change in system pressure.

(a) The constraint here is the temperature and if the temperature is increased, the system will behave as if trying to lower it by favoring endothermic reaction in the system i.e. favoring the direction of the equilibrium which absorbs heat. The reverse is the case when the temperature is lowered.

(b) The constraint here is concentration. If the concentration of one of the components of the system is increased, the system will adjust by shifting the equilibrium to a state which favors the consumption of the added component. If on the other hand a component of the system is removed, the equilibrium position will shift in an attempt to replenish some of the removed component.

(c) The constraint here is pressure. In gaseous reactions, or systems which is compressible, increase in the pressure or otherwise will introduce some constraint to the system. However, the system will adjust itself to counter the effect of such constraint. In gaseous systems, the increase of pressure is similar to increasing the concentration of the components and decrease is similar to decreasing the concentration.

Factors Affecting the Rate of Chemical Reaction

1. Temperature

2. Accessibility of reactants

3. Concentration

4. Presence of catalysts.

A catalyst is a substance which can increase the rate of a reaction without itself undergoing permanent chemical change.

Properties of a Catalyst

1. A small quantity (number of moles) of catalyst can catalyze a very large number of moles of reactant.

2. A catalyst may undergo intermediate physical changes and may form temporary chemical bonds with the reactant but it will be unchanged in amount and composition at the end of the reaction.

3. A catalyst cannot affect the (free energy) G of a reaction and hence, it cannot change the equilibrium position but may accelerate the attainment of equilibrium.

4. The function of a catalyst can be likened to "Lubricating the mechanism". It changes the rate of a reaction but not its equilibrium.

Free Energy change ΔG

The energy change that takes place under equilibrium condition.

Entropy change of a reaction ΔS

The degree of disorderliness of a system is measured by - a quantity called entropy. **+ve S** implies that the system is more disordered. **-ve S** implies that the system is less disordered.

Enthalpy change ΔH

Enthalpy change is the heat change of a process at constant pressure.

Heat of formation is defined as the quantity of heat that is evolved or absorbed when one mole of a compound is formed from its elements in their standard states. It is denoted by $\Delta H°_f$

Exothermic reaction is a reaction which evolves heat. A compound formed exothermically from its reactants is called an exothermic compound. An energy diagram of an exothermic reaction shows that the products' energy levels are below those of reactants.

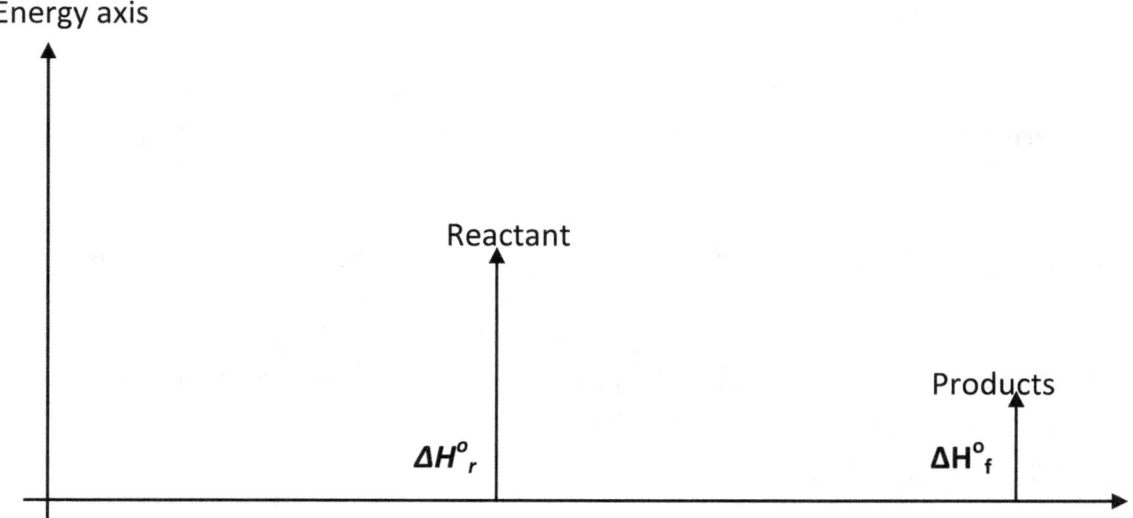

Endothermic reaction is a reaction which absorbs heat or that takes place in a heated environment. Such a product is called an endothermic compound. An endothermic compound (product) can be regarded as having a large amount of stored energy and is above the reactant level in the energy diagram.

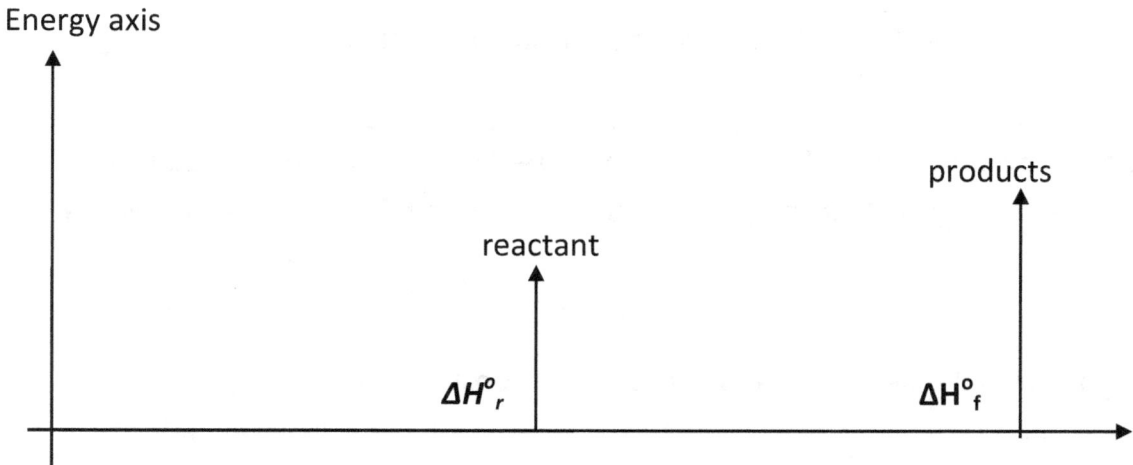

Thermoneutral reactions: A reaction which neither evolves nor absorbs heat. Both the reactants and product are on the same level in the energy diagram.

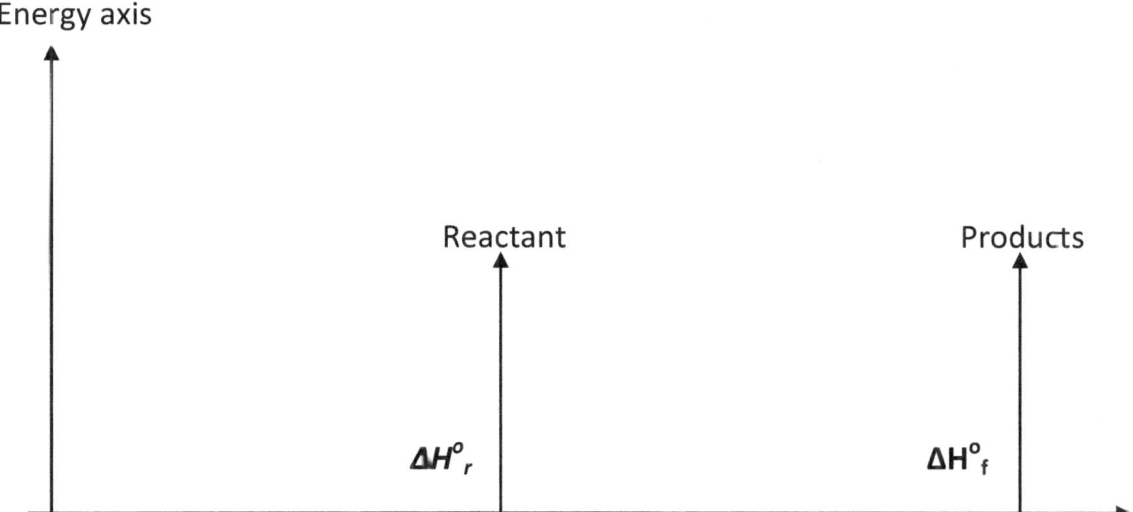

CONSTANTS

Density: **d = m/V (g/cm³)**

Where m is the mass of a substance and V is its volume

Charles's Law: **V = kT**

Where V is volume, k is Boltzmann's constant, and T is the temperature in Kelvin

Ideal Gas Equation: **PV = nRT**

Where P is pressure, V is volume, n is the number of moles of gas, R is the gas constant, and T is the temperature in Kelvin

Molarity: **M = n/V**

Where n is the number of moles of solute and V is the volume of solution in liters

Molarity equation: **$M_1V_1 = M_2V_2$**

Where M_1 is the initial molarity of a solution, V_1 is its initial volume, M_2 is its diluted molarity, and V_2 is its diluted volume

Molality: **M = n/ms**

Where n is the number of moles of solute and ms is the mass of solvent in kilograms

pH: **pH = -log(H⁺)**

Where *[H⁺]* is the concentration of hydronium ions *(H_3O^+)* in the solution

Avogadro's constant: **N_A = 6.022142 × 10²³ mol⁻¹**

Gas constant: **R = 8.31447 J/mol·K**

Boltzmann constant: **k = 1.3806503 × 10⁻²³ J/K**

Planck constant: **h = 6.6260688 × 10⁻³⁴ J/s**

Elementary charge: **e = 1.602176 × 10⁻¹⁹ C**

Electron rest mass: **me = 9.10938189 × 10⁻³¹ kg**

Proton rest mass: **mp = 1.67262158 × 10⁻²⁷ kg**

Neutron rest mass: **mn = 1.6749286 × 10⁻²⁷ kg**

Practice questions for Students C7

1. What do you understand by the term activity series?

2. (a) What is electronegativity? Explain in detail.

   (b) What is an element? Explain in detail

   (c) What is an electron? Explain in detail.

3. Explain Boyle's law from first principle.

APPENDIX I

Table of Oxidation Number of some common elements

Element	Symbol	Atomic No	Possible oxidation Nos
Hydrogen	H	1	+1, -1
Helium	He	2	
Lithium	Li	3	+1
Beryllium	Be	4	+2
Boron	B	5	+3
Carbon	C	6	+4, +2, -4
Nitrogen	N	7	+5, +4, +3, +2, +1, -3
Oxygen	O	8	-1, -2
Fluorine	F	9	-1
Neon	Ne	10	
Sodium	Na	11	+1
Magnesium	Mg	12	+2
Aluminum	Al	13	+3
Silicon	Si	14	+4, -4
Phosphorus	P	15	+5, +3, -3
Sulphur	S	16	+6, +4, +2, -2
Chlorine	Cl	17	-7, +5, +3, +1, -1
Argon	Ar	18	
Potassium	K	19	+1
Calcium	Ca	20	+2
Scandium	Sc	21	+3

Vanadium	V	23	+5, +4, +3, +2
Chromium	Cr	24	+6, +3, +2
Manganese	Mn	25	+7, +6, +4, +3, +2
Iron	Fe	26	+3, +2
Cobalt	Co	27	+3, +2
Nickel	Ni	28	−1, +1, −2, +3, +4
Copper	Cu	29	+2, +1
Zinc	Zn	30	+2
Gallium	Ga	31	+3
Germanium	Ge	32	+4, −4
Arsenic	As	33	+5, +3, −3
Selenium	Se	34	+6, +4, −2
Bromine	Br	35	+5, +1, −1
Krypton	Kr	36	+4, +2
Palladium	Pd	46	+4, +2
Silver	Ag	47	+1
Cadmium	Cd	48	+2
Indium	In	49	+3
Tin	Sn	50	−4, +2
Antimony	Sb	51	+5, +3, −3
Iodine	I	53	+7, +5, +1, −1
Xenon	Xe	54	+6, −4, +2
Cesium	Cs	55	+1

Tungsten	W	74	+5, +4
Platinum	Pt	78	+4, +2
Gold	Au	79	+3, +1
Mercury	Hg	80	+2, +1
Lead	Pb	82	+4, +2
Bismuth	Bi	83	+5, +3
Barium	Ba	56	+2, +4

APPENDIX II

CHEMICAL ELEMENTS SORTED IN ALPHABETICAL ORDER WITH SOME FACTS ABOUT THEIR DISCOVERY

Name	Symbol	Atomic No.	Atomic Weight	Group	Date Discovered	Discovered By
Actinium	Ac	89	(227)	Actinide Series	1899	André Debierne
Aluminum	Al	13	26.9815	Other Metals	1824	Hans Oersted (also attributed to Friedrich Wöhler 1827)
Americium	Am	95	243	Actinide series	1944	Glenn Seaborg, Ralph James, Leon Morgan, and Albert Ghiorso
Antimony	Sb	51	121.760	Other metals	prehistoric	unknown
Argon	Ar	18	39.948	Noble gases	1894	John Rayleigh and Ramsay William
Arsenic	As	33	74.9216	Non metals	prehistoric	unknown
Astatine	At	85	(210)	Halogen	1940	Dale R. Corson, K. R. MacKenzie and Emilo Segrè
Barium	Ba	56	137.328	Alkaline earth metals	1808	Humphry Davy
Berkelium	Bk	97	(247)	Actinide series	1949	Glenn Seaborg, Stanley Albert Thompson, and Ghiorso
Beryllium	Be	4	9.0122	Alkaline earth metals	1798	Louis-Nicolas Vauquelin (isolated by Friedrich Wöhler and Antoine-Alexandre-Brutus Bussy 1828)
Bismuth	Bi	83	208.9804	Other metals	prehistoric	unknown
Bohrium	Bh		(262)	Transition metals	1976	Georgii Flerov and Yuri Oganessian (confirmed by German scientist Peter Armbruster and coworkers)
Boron	B	5	10.81	Non metals	1808	Humphry Davy, and independently by Joseph Gay-Lussac and Louis-Jacques Thénard
Bromine	Br	35	79.904	Halogens	1826	Antoine-Jérôme Balard

Element	Symbol	Atomic #	Atomic Mass	Category	Year	Discoverer
Cadmium	Cd	48	112.412	Transition metals	1817	Friedrich Strohmeyer
Calcium	Ca	20	40.078	Alkaline earth metals	1808	Humphry Davy
Californium	Cf	98	(251)	Actinide series	1950	Glenn Seaborg, Stanley Thompson, Kenneth Street, Jr., and Albert Ghiorso
Carbon	C	6	12.011	Non metals	prehistoric	unknown
Cerium	Ce	58	140.115	Lanthanide series	1804	Jöns Berzelius and Wilhelm Hisinger and independently by Martin Klaproth
Cesium	Cs	55	132.9054	Alkali metals	1860	Robert Bunsen and Gustav Kirchhoff
Chlorine	Cl	17	35.4528	Halogen	1774	Karl Scheele
Chromium	Cr	24	51.9962	Transition metals	1797	Louis-Nicolas Vauquelin
Cobalt	Co	27	58.9332	Transition metals	1730	Georg Brandt
Copper	Cu	29	63.546	Transition metals	prehistoric	unknown
Curium	Cm	96	(247)	Actinide series	1944	Glenn Seaborg, Ralph and Albert Ghiorso James
Darmstadtium	Ds	110	(271)	Transition metals	1994	team at the Heavy-Ion Research Laboratory, Darmstadt, Germany
Dubnium	Db	105	(262)	Transition metals	1970	Claimed by U.S. Scientist Albert Ghiorso and coworkers (disputed by Soviet workers)
Dysprosium	Dy	66	162.500	Lanthanide series	1886	Paul Lecoq de Boisbaudran Soviet workers)
Einsteinium	Es	99	(252)	Actinide series	1952	Albert Ghiorso and coworkers
Erbium	Er	68	167.26	Lanthanide series	1843	Carl Mosander
Europium	Eu	63	151.966	Lanthanide series	1901	Eugène Demarçay
Fermium	Fm	100	(257)	Actinide series	1955	Albert Ghiorso and coworkers
Fluorine	F	9	18.9984	Halogens	1771	Karl Scheele (isolated by Henri Moissan 1886)
Francium	Fr	87	(223)	Alkali metals	1939	Marguérite Perey
Gadolinium	Gd	64	157.25	Lanthanide series	1886	Paul Lecoq de Boisbaudran
Gallium	Ga	31	69.723	Other metals	1875	Paul Lecoq de Boisbaudran
Germanium	Ge	32	72.61	Other metals	1886	Clemens Winkler

Gold	Au	79	196.9665	Transition metals	Prehistoric	unknown
Hafnium	Hf	72	178.49	Transition metals	1913	Dirk Coster and Georg von Hevesy
Hassium	Hs	108	(263)	Transition metals	1984	Peter Armbruster and coworkers
Helium	He	2	4.0026	Noble gases	1868	Pierre Janssen
Holmium	Ho	67	164.9303	Lanthanide series	1879	Per Cleve
Hydrogen	H	1	1.0079	Nonmetals	1766	Henry Cavendish
Indium	In	49	114.818	Other metals	1863	Ferdinand Reich and Hieronymus Richter
Iodine	I	53	126.9045	Halogens	1811	Bernard Courtois
Iridium	Ir	77	192.217	Transition metals	1804	Smithson Tennant
Iron	Fe	26	55.845	Transition metals	Prehistoric	unknown
Krypton	Kr	36	83.798	Noble gases	1898	William Ramsay and Morris Travers
Lanthanum	La	57	138.9055	Lanthanide series	1839	Carl Mosander
Lawrencium	Lr	103	(260)	Transition metals	1961	Albert Ghiorso, Torbjørn Sikkeland, Almon Larsh, and Robert Latimer
Lead	Pb	82	207.2	Other metals	Prehistoric	unknown
Lithium	Li	3	6.941	Alkali metals	1817	Johan Arfwedson
Lutetium	Lu	71	174.967	Transition metals	1907	Georges Urbain and Carl von Welsbach, independently of each other
Magnesium	Mg	12	24.3051	Alkaline earth metals	1755	Joseph Black (oxide isolated by Humphry Davy 1808; pure form isolated by Antoine-Alexandre-Brutus Bussy 1828)
Manganese	Mn	25	54.938	Transition metals	1774	Johann Gottlieb Gahn
Meitnerium	Mt	109	(268)	Transition metals	1982	Peter Armbruster and coworkers
Mendelevium	Md	101	(258)	Actinide series	1955	Albert Ghiorso, Bernard G. Harvey, Gregory Choppin, Stanley Thompson, and Glenn Seaborg
Mercury	Hg	80	200.59	Transition metals	prehistoric	unknown
Molybdenum	Mo	42	95.94	Transition metals	1781	named by Karl Scheele (isolated by Peter Jacob Hjelm 1782)

Neodymium	Nd	60	144.24	Lanthanide series	1885	Carl von Welsbach
Neon	Ne	10	20.1798	Noble gases	1898	William Ramsay and Morris Travers
Neptunium	Np	93	(237)	Actinide series	1940	Edwin McMillan and
Nickel	Ni	28	58.6934	Transition metals	1751	Axel Cronstedt
Niobium	Nb	41	92.9064	Transition metals	1801	Charles Hatchett
Nitrogen	N	7	14.0067	Nonmetals	1772	Daniel Rutherford
Nobelium	No	102	(259)	Actinide series	1958	Albert Ghiorso, Torbjørn Sikkeland, J. R. Walton, and Glenn Seaborg
Osmium	Os	76	190.23	Transition metals	1804	Smithson Tennant
Oxygen	O	8	15.9994	Nonmetals	1774	Joseph Priestley and Karl Scheele, independently of each other
Palladium	Pd	46	106.42	Transition metals	1804	William Wollaston
Phosphorus	P	15	30.9738	Nonmetals	1674	Hennig Brand
Platinum	Pt	78	195.08	Transition metals	1557	Julius Scaliger
Plutonium	Pu	94	(244)	Actinide series	1940	Glenn Seaborg, Edwin McMillan, Joseph Kennedy, and Arthur Wahl
Polonium	Po	84	(209)	Other metals	1898	Marie and Pierre Curie
Potassium	K	19	39.0983	Alkali metals	1807	Humphry Davy
Praseodymium	Pr	59	140.908	Lanthanide series	1885	Carl von Welsbach
Promethium	Pm	61	(145)	Lanthanide series	1945	J. A. Marinsky, Lawrence Glendenin, and Charles Coryell
Protactinium	Pa	91	231.036	Actinide series	1913	Kasimir Fajans and O. Göhring
Radium	Ra	88	(226)	Alkaline earth metals	1898	Marie Curie
Radon	Rn	86	(222)	Noble gases	1900	Friedrich Dorn
Rhenium	Re	75	86.207	Transition metals	1925	Walter Noddack, Ida Tacke, and Otto Berg
Rhodium	Rh	45	102.9055	Transition metals	1804	William Wollaston
Rubidium	Rb	37	85.4678	Alkali metals	1861	Robert Bunsen and Gustav Kirchhoff

Name	Symbol	Atomic #	Atomic Mass	Category	Year	Discoverer
Ruthenium	Ru	44	101.07	Transition metals	1827	G. W. Osann (isolated by Karl Klaus 1844)
Rutherfordium	Rf	104	(261)	Transition metals	1969	Claimed by U.S. Scientist Albert Ghiorso and coworkers (disputed by Soviet workers)
Roentgenium	Rg	111	(272)	Transition metals	1994	Team at the Heavy-Ion Research Laboratory, Darmstact, Germany
Samarium	Sm	62	150.36	Lanthanide series	1879	Paul Lecoq de Boisbaudran
Scandium	Sc	21	44.9559	Transition metals	1876	Lars Nilson
Seaborgium	Sg	106	(266)	Transition metals	1974	claimed by Georgi Flerov and coworkers, and independently by Albert Ghiorso and coworkers
Selenium	Se	34	78.96	Nonmetals	1817	Jöns Berzelius
Silicon	Si	14	28.0855	Nonmetals	1823	Johan Arfwedson
Silver	Ag	47	107.8682	Transition metals	prehistoric	unknown
Sodium	Na	11	22.9898	Alkali metals	1807	Humphry Davy
Strontium	Sr	38	87.62	Alkaline earth metals	1808	Humphry Davy
Sulfur	S	16	32.067	Nonmetals	Prehistoric	unknown
Tantalum	Ta	73	180.948	Transition metals	1802	Anders Ekeberg
Technetium	Tc	43	(98)	Transition metals	1937	Carlo Perrier and Emilio Segrè
Tellurium	Te	52	127.60	Nonmetals	1782	Franz Müller
Terbium	Tb	65	158.9253	Lanthanide series	1843	Carl Mosander
Thallium	Tl	81	204.3833	Other metals	1861	William Crookes (isolated by William Crookes and Claude August Lamy, independently of each other, in 1862)
Thorium	Th	90	232.0381	Actinide series	1828	Jöns Berzelius
Thulium	Tm	69	158.9342	Lanthanide series	1879	Per Cleve
Tin	Sn	50	118.711	Other metals	Prehistoric	unknown
Titanium	Ti	22	47.867	Transition metals	1790	William Gregor
Tungsten	W	74	183.84	Transition metals	1783	isolated by Juan José Elhuyar

Name	Symbol	Z	Atomic Mass	Category	Year	Discoverer
Ununbium	Uub	112	(277)	Transition metals	1996	team at the Heavy-Ion Research Laboratory, Darmstadt, Germany
Ununhexium	Uuh	116	(292)	Other metals	2000	team at the Joint Institute for Nuclear Research, Dubna, Russia
Ununoctium	Uuo	118	(294)	Noble gases	2006	team at the Joint Institute For Nuclear Research, Dubna, Russia
Ununpentium	Uup	115	(288)	Other metals	2004	team at the Joint Institute For Nuclear Research, Dubna, Russia
Ununquadium	Uuq	114	(285)	Other metals	1998	team at the Joint Institute for Nuclear Research, Dubna, Russia
Ununtrium	Uut	113	(284)	Other metals	2004	team at the Joint Institute for Nuclear Research, Dubna, Russia
Uranium	U	92	238.0289	Actinide series	1789	Martin Klaproth (isolated by Eugène Péligot 1841)
Vanadium	V	23	50.9415	Transition metals	1801	Andrés del Rio (disputed), or Nils Sefström 1830
Xenon	Xe	54	131.29	Noble gases	1898	William Ramsay and Morris Travers
Ytterbium	Yb	70	173.04	Lanthanide series	1878	Jean Charles de Marignac
Yttrium	Y	39	88.906	Transition metals	1794	Johan Gadolin
Zinc	Zn	30	65.409	Transition metals	prehistoric	unknown
Zirconium	Zr	40	91.224	Transition metals	1789	Martin Klaproth

APPENDIX III

CHEMICAL ELEMENTS SORTED IN ALPHABETICAL ORDER WITH SOME INTERESTING FACTS ABOUT THEM.

Symbol	Atomic No.	Name	Interesting Facts
Ac	89	Actinium	Radioactive metal; rare
Al	13	Aluminum	Alloys are strong and light weight
Am	95	Americium	Made by man; highly radioactive
Sb	51	Antimony	Silvery, brittle metal, important alloy metal
Ar	18	Argon	Colorless gas in air; used in electric light bulbs
As	33	Arsenic	Grey solid; compounds are poisonous
At	85	Astatine	Made by man from bismuth; radioactive non-mental
Ba	56	Barium	Lightweight metal; soft silver white
Bk	97	Berkelium	Made by man (1950); highly radioactive metal
Be	4	Beryllium	Lightweight metal; small beryllium-copper springs are very long lived
Bi	83	Bismuth	Silvery pink metal; makes hard alloys with low melting point
B	5	Boron	Solid non-metal; present in borax
Br	35	Bromine	Red liquid; name means stretch
Cd	48	Cadmium	Silvery metal; often electroplated over other metals
Ca	20	Calcium	Light weight metal, compound are abundant in earth's crust
Cf	98	Californium	Made by man (1950); highly radioactive metal
C	6	Carbon	Key element of organic chemistry in all plants and animals
Ce	58	Cerium	Hard metal; makes sparks for lighters
Cs	55	Cesium	Soft, silvery metal; melts in boiling water
Cl	17	Chlorine	Greenish-yellow gas; poisonous; good bleaching agent
Cr	24	Chromium	Bright, silvery metal; used in stainless steel Alloys
Co	27	Cobalt	Silvery metal; part of powerful magnetic alloy
Cu	29	Copper	Red metal; conductor of electricity
Cm	96	Curium	Made by man from plutonium; highly radioactive metal
Dy	66	Dysprosium	Rare earth metal; name means 'hard to get at'
Es	99	Einsteinium	Made by bombarding uranium with the nuclei of nitrogen atoms; atomic weight 247; highly radioactive
Er	68	Erbium	Rare earth metal
Eu	63	Europium	Rare earth metal
Fm	100	Fermium	Made by adding neutrons to plutonium to make californium and then more neutrons to make element 100; highly

				radioactive
F	9	Fluorine		Highly active and poisonous gas
Fr	87	Francium		Radioactive metal; extremely rare; also produced by nuclear reactions
Gd	64	Gadolinium		Rare earth metal; free element not yet prepared
Ga	31	Gallium		Shining, white metal; usually separated from zinc ores
Ge	32	Germanium		Grey brittle metal; similar to tin
Au	79	Gold		Long famous for decoration and money standard
Hf	72	Hafnium		Heavy metal; similar to zirconium
He	2	Helium		Chemically inactive gas twice as heavy as hydrogen
Ho	67	Holmium		Rare earth metal; free element not yet produced
H	1	Hydrogen		Colourless gas, lightest of all
In	49	Indium		Soft, silvery metal; similar to aluminum
I	53	Iodine		Brownish-black solid; when heated Changes to beautiful purple vapor
Ir	77	Iridium		Silvery metal; alloyed with platinum for pen points
Fe	26	Iron		Second most abundant metal
Kr	36	Krypton		Inert colourless gas in air
La	57	Lanthanum		Rare earth metal
Lw	103	Lawrencium		Newest element, extremely short life, radioactive, manmade (1961)
Pb	82	Lead		Heavy, bluish-white, soft metal
Li	3	Lithium		Lightest metal known; soft
Lu	71	Lutetium		Rare earth metal; of little use
Mg	12	Magnesium		Combines lightness with strength
Mn	25	Manganese		Heavy metal; highly important in steel industry
Md	101	Mendelevium		Short lived, highly radioactive
Hg	80	Mercury		Heavy, silvery, liquid metal
Mo	42	Molybdenum		Silvery metal; has many important steel alloys
Nd	60	Neodymium		Rare earth metal; compounds pink
Ne	10	Neon		Inert gas in the air; makes brilliant electric lights
Np	93	Neptunium		Made by man from uranium; radioactive
Ni	28	Nickel		Makes tough corrosion-resistant steel
Nb	41	Niobium		Silvery metal; formerly called columbium
N	7	Nitrogen		Colorless gas; makes up78% of air
No	102	Nobelium		Short lived; highly radioactive
Os	76	Osmium		Silvery metal; heaviest element
O	8	Oxygen		Colourless gas; abundant element
Pd	46	Palladium		Resembles platinum
P	15	Phosphorous		Soft, non-metallic solid; ignites easily
Pt	78	Platinum		Silvery metal; useful for laboratory vessels and instruments
Pu	94	Plutonium		Made by man; highly important for nuclear fission
Po	84	Polonium		Radioactive metal; found by the Curies just before radium
K	19	Potassium		Soft metal; lighter than water

Pr	59	Praseodymium	Rare earth metal
Pm	61	Promethium	Rare earth metal; also made by man from praseodymium
Pa	91	Protactinium	Radioactive metal; present in all uranium ores
Ra	88	Radium	Radioactive metal; discovery stimulated research on radioactive
Rn	86	Radon	Heaviest gas known; radioactive; comes from radium
Re	75	Rhenium	Heavy metal; resembles manganese
Rh	45	Rhodium	Heavy metal; looks like aluminum; used for electroplating jewelry
Rb	37	Rubidium	Soft metal; rare; highly active chemically
Ru	44	Ruthenium	Hard, grey, brittle metal
Sm	62	Samarium	Rare earth metal
Sc	21	Scandium	Free element not yet produced; rare
Se	34	Selenium	Solid non-metal; resembles sulphur in chemical changes
Si	14	Silicon	Solid non-metal; second in abundance
Ag	47	Silver	Best conductor of heat and electricity
Na	11	Sodium	Soft, highly active metal; lighter than water
Sr	38	Strontium	Hard, active metal; resembles Calcium chemically
S	16	Sulphur	Solid yellow non - metal
Ta	73	Tantalum	Looks like polished iron; makes alloy steel
Tc	43	Technetium	Heavy metal; found among fission products of uranium
Te	52	Tellurium	Solid non-mental; resembles sulphur in chemical changes
Tb	65	Terbium	Rare earth metal; free metal not yet prepared
Ti	81	Thallium	Solid metal; resembles lead; its salts are very poisonous
Th	90	Thorium	Heavy, grey metal; all its compounds are radioactive
Tm	69	Thulium	Rare earth metal; not yet obtained as free element
Sn	50	Tin	Silvery metal; electroplated on steel for cans
Ti	22	Titanium	Strong, hard metal; new production methods makes future bright
U	92	Uranium	Object of world-wide search because of need for nuclear fission
V	23	Vanadium	Grey metal; difficult to melt; makes strong, tough alloy of steel
W	74	Tungsten	Heavy metal; formerly called wolfram; has highest melting point
Xe	54	Xenon	Rare, inert, colored gas in the air
Yb	70	Ytterbium	Rare earth metal
Y	39	Yttrium	Rare earth metal; more abundant than many other rare earth metals
Zn	30	Zinc	Bluish-white metal; used for other coat on galvanized iron
Zr	40	Zirconium	Gold-coloured metal; the compound, zircon is a gem

APPENDIX IV

CHECK LISTS OF CHEMICAL FORMULAS AND THEIR IUPAC NOMENCLATURE

Formular   **Name**

[A]

AgCl   Silver chloride

AgNO$_3$ Silver trioxo nitrate(V)

Al$_4$C$_3$ Aluminium Carbide

Al$_2$O$_3$ Aluminium (III) oxide

As$_2$O$_3$ Arsenic (III) oxide

Au$_2$O$_3$ Gold III Oxide

Au$_2$(NO$_3$)$_6$ Gold III trioxo nitrate (V)

[B]

BaO   Barium(II) oxide

BaO$_2$ Barium(II) peroxide

BaSO$_4$ Barium(II) tetraoxo sulphate(VI)

BiOCl   Bismuth(III) chloride oxide

Br$^-$ Bromide ion

[C]

Cl$^-$ Chloride ion

CaC$_2$ Calcium dicarbide

CaCl$_2$ Calcium dichloride

CaCO$_3$ Calcium trioxo carbonate(IV)

Ca(HCO$_3$)$_2$ Calcium hydrogen trioxo carbonate (IV)

CaO	Calcium oxide
$CaSO_4$	Calcium tetraoxo sulphate(VI)
$Ca(OH)_2$	Calcium hydroxide
CBr_4	Tetrabromomethane
CCl_4	Tetrachloromethane
Cf_4	Tetrafluoromethane
$CHCl_3$	Trichloromethane
$CHBr_3$	Tribromomethane
CH_2Cl_2	Dichloromethane
CI_4	Tetraiodomethane
CO	Carbon(II) oxide
CO_2	Carbon(IV) oxide
CO_3^{2-}	Trioxo carbonate(IV) ion
CrO	Chromium(II) oxide
Cr_2O_3	Chromium(III) oxide
CrO_3	Chromium(VI) oxide
Cr_2S_3	Chromium(III) sulphide
$CrSO_4$	Chromium(II) tetraoxo sulphate(VI)
$CuCl_2$	Copper(II) chloride
$Cu_2CO_3(OH)_2$	Dicopper(II)trioxo carbonate(IV) dihydroxide
CS_2	Carbon(IV) sulphide
CuO	Copper(II) oxide
Cu_2O	Copper(I) oxide
$Cu(NH_3)_4Cl_2$	Tetraammine copper(II) dichloride
$Cu(OH)_2$	Copper(II) hydroxide

Cu₂(OH)₃Cl	Copper(II) Chloride trihydroxide
CuSO₄	Copper(II) tetraoxo sulphate(VI)
CuSO₄.5H₂O	Copper(II) tetraoxo sulphate(VI) pentahydrate

[F]

F⁻	Fluoride ion
FeCl₂	Iron(II) chloride
FeCl₃	Iron(III) chloride
FeO	Iron(II) oxide
Fe₂O₃	Iron(III) oxide
Fe₃O₄	Triiron tetraoxide (mixoxide) (i.e iron(II) iron(III) Oxide)
Fe(OH)₂	Iron(II) hydroxide
Fe(OH)₃	Iron(III) hydroxide
FeS	Iron(II) sulphide
FeSO₄	Iron(II) tetraoxo sulphate(VI)
Fe₂SO₄.7H₂O	Iron(II) tetraoxo sulphate(VI)-7-water
Fe₂(SO₄)₃	Iron(III) tetraoxo sulphate

[H]

H⁺	Hydrogen ion (proton)
H⁻	Hydride ion
HBr	Hydrogen bromide acid
HCl	Hydrogen chloride acid
HgCl	Mercury(I) chloride
HgCl₂	Mercury(II) chloride
HF	Hydrogen Fluoride acid

HI	Hydrogen iodide acid
H_2O	Hydrogen Oxide (water)
H_2CO_3	Trioxo carbonate(IV) acid
HNO_3	Trioxo nitrate(V) acid
H_3PO_4	Tetraoxo phospate(V) acid
H_2SO_4	Tetraoxo sulphate(VI) acid
H_2SO_3	Trioxo sulphate(IV) acid

[I]

I^-	Iodide ion
ICl	Iodine(I) chloride
I_2O_5	Iodine(V) oxide

[K]

K^+	Potassium ion
$KAl(SO_4)_2 \cdot 12H_2O$	Aluminium potassium bi(tetraoxo sulphate(VI)) dodecahydrate
KBr	Potassium bromide
$K_2Cr_2O_7$	Potassium heptaoxo dichromate(VI)
KClO	Potassium monoxo chlorate(II)
$KClO_3$	Potassium trioxo chlorate(V)
$K_4Fe(CN)_6$	Tetra Potassium hexacyano Iron(II)
$K_3Fe(CN)_6$	Tri Potassium hexacyano ferrate(III)
KI	Potassium iodide
$KMnO_4$	Potassium tetraoxo manganate(VII)
$K_4Ni(CN)_4$	Tetra Potassium tetracyano Nicollate(O)
K_2O	Potassium oxide

K$_2$SO$_4$ — Di Potassium tetraoxo sulphate(VI)

KAg(CN)$_2$ — Potassium dicyanosilver

[M]

Mg^{2+} — Magnesium ion

MgCl$_2$ — Magnesium chloride

MgO — Magnesium oxide

Mg(OH)$_2$ — Magnesium hydroxide

Mg(OH)Cl — Magnesium chloride hydroxide

MgSO$_4$ — Magnesium tetraoxo sulphate(VI)

MgSO$_4$.7H$_2$O — Magnesium tetraoxo sulphate(VI)-7-water

Mn^{2+} — Manganese ion

MnCl$_2$ — Manganese(II) chloride

MnO — Manganese(II) oxide

MnO$_2$ — Manganese(IV) oxide

Mn2O$_3$ — Manganese(III) oxide

Mn$_3$O$_4$ — Trimanganese tetraoxide [mixed Oxide of Mn(II) and Mn(III) Oxide

MnSO$_4$ — Manganese(II) tetraoxo sulphate(VI)

[N]

Na$^+$ — Sodium ion

NaCl — Sodium chloride

NaClO — Sodium monoxo chlorate(I)

NaClO$_3$ — Sodium trioxo chlorate(V)

Na$_2$CO$_3$ — Sodium trioxo carbonate(IV)

NaH — Sodium hydride

NaHCO₃	Sodium hydrogen trioxo carbonate(IV)
NaH₂PO₄	Sodium dihydrogen tetraoxo phosphate (V)
NaHSO₃	Sodium hydrogen trioxo sulphate(IV)
NaHSO₄	Sodium hydrogen tetraoxo sulphate (VI)
NO²⁻	Dioxonitrate(III) ion
NO³⁻	Trioxonitrate(V) ion
NaOH	Sodium hydroxide
Na₂SO₄	Di Sodium tetraoxo sulphate(VI)
Na₂SO₃	Di Sodium trioxo sulphate(IV)
NH₃	Ammonia molecule
NH₄⁺	Ammonium ion
N₂H₄	Hydrazine Molecule
NH₄OH	Ammonia solution(or ammonium hydroxide)
NH₄NO₂	Ammonium dioxo nitriate(III)
NH₄NO₃	Ammonium trioxo nitriate(V)
NH₄Al(SO₄)₂	Aluminium ammonium tetraoxo sulphate(VI)
(NH₄)₂Fe(SO₄)₂	Ammorium iron (II) tetraoxo sulphate (VI)
N₂O	Nitrogen(I) oxide
NO	Nitrogen(II) oxide
NO₂	Nitrogen(IV) oxide
N₂O₄	Dinitrogen tetraoxide (mix oxide)
N₂O₅	Nitrogen(V) oxide
NO₃	Nitrogen(VI) oxide
NiO	Nickel(II) oxide
NiO₂	Nickel(IV) oxide

Ni(OH)$_2$ — Nickel(II) hydroxide

(NH$_4$)$_2$ Ni(SO$_4$)$_2$ — Nickel (II) Ammonium tetraoxo sulphate (VI)

[P]

Pb^{2+} — Lead(II) ion

PbCl$_3$ — Lead(III) chloride

PbCl$_5$ — Lead(V) chloride

PO$_4^{2-}$ — Tetraoxo phosphate(VI) ion

PO$_4^{3-}$ — Tetraoxo phosphate(V) ion

P$_4$O$_{10}$ — Tetraphosphorus decaoxide

PbO — Lead(II) oxide

PbO$_2$ — Lead(IV) oxide

Pb$_3$O$_4$ — Trilead tetraoxide (mixed Oxide)

PbS — Lead(II) sulphide

PbSO$_3$ — Lead(II) trioxo sulphate(IV)

PbSO$_4$ — Lead(II) tetraoxo sulphate(VI)

Pb(NO$_3$)$_2$ — Lead(II) trioxo nitrate(V)

PCl$_3$ — Phosphorous(III) chloride

PCl$_5$ — Phosphorous(V) chloride

[S]

Sb$_2$O$_3$ — Antimony(III) oxide

SiO$_2$ — Silicon(IV) oxide

SO$_2$ — Sulphur(IV) oxide

SO$_3$ — Sulphur(VI) oxide

[Z]

ZnO Zinc oxide

ZnCl$_2$ Zinc chloride

ZnSO$_4$ Zinc tetraoxo sulphate(VI)

APPENDIX V

A CHECK LIST OF BALANCED CHEMICAL EQUATIONS

1. DECOMPOSITION REACTION EQUATIONS

$Ca(HCO_3)_2 \longrightarrow CaCO_3 + H_2O + CO_2$

$2KClO_{3(s)} \longrightarrow 2KCl_{(s)} + 3O_{2(g)}$

$2Al(OH)_3 \longrightarrow Al_2O_3 + 3H_2O$

$2ClO^-_{3(s)} \longrightarrow 2Cl^-_{(s)} + 3O_2$

$2Zn(NO_3)_{2(s)} \longrightarrow 2ZnO_{(s)} + 4NO_2 + O_{2(g)}$

$2FeSO_{4(s)} \longrightarrow Fe_2O_{3(s)} + SO_2 + SO_3$

$2HgO_{(s)} \longrightarrow 2Hg + O_{2(g)}$

$2PbO_{2(s)} \longrightarrow 2PbO_{(s)} + O_{2(g)}$

$2PbO + 2PbO_2 \longrightarrow 2Pb_2O_3$

$2Pb_2O_3 \longrightarrow 2PbO + 2PbO_2$

$2Pb_3O_{4(s)} \longleftrightarrow 6PbO_{(s)} + O_{2(g)}$

$H_2S_{(g)} \longrightarrow H_{2(g)} + S_{(s)}$

$ZnCl_2 \cdot H_2O_{(s)} \xrightarrow{\Delta H} Zn(OH)Cl + HCl_{(g)}$

$2NaHCO_{3(s)} \longrightarrow Na_2CO_{3(s)} + H_2O_{(l)} + CO_{2(g)}$

$CaCO_{3(s)} \longrightarrow CaO_{(s)} + CO_{2(g)}$

$NH_4Cl_{(s)} \longleftrightarrow NH_{3(g)} + HCl_{(g)}$

$2NH_4VO_{3(s)} \longrightarrow V_2O_5 + 2NH_3 + H_2O$

$FeCr_2O_{4(s)} + 4C \longrightarrow Fe + 2Cr + 4CO_{(g)}$

$3MnO_{2(s)} \longrightarrow Mn_3O_4 + O_2$

$2Mn_2O_{7(s)} \longrightarrow 4MnO_2 + 3O_{2(g)}$

$Mn(NO_3)_{2(s)} \longrightarrow MnO_2 + 2NO_2$

$MnC_2O_{4(s)} \longrightarrow MnO + CO + CO_2$

$2HCl + CaCO_3 \xrightarrow{\text{Double decomposition}} CaCl_2 + H_2O + CO_2$

$FeC_2O_{4(s)} \longrightarrow FeO + CO + CO_2$

$4FeO_{(s)} \longrightarrow Fe + Fe_3O_4$

$CoCO_{3(s)} \longrightarrow CoO + CO_2$

$2Co(NO_3)_{2(s)} \longrightarrow 2CoO + 4NO_2 + O_2$

$Ni(CO)_{4(s)} \xrightarrow{200^0C} Ni + 4CO$

$NiCO_{3(s)} \longrightarrow NiO + CO_2$

$2Ni(NO_3)_2 \longrightarrow 2NiO + 4NO_2 + O_2$

2. ADDITION REACTION EQUATIONS

$ZnO_{(s)} + 2NaOH_{(aq)} + H_2O_{(l)} \longrightarrow Na_2Zn(OH)_{4(aq)}$

$Al_2O_{3(s)} + NaOH_{(aq)} + 3H_2O_{(l)} \longrightarrow NaAl(OH)_{4(aq)} + Al(OH)_3$

$PbO + NaOH_{(aq)} + H_2O_{(l)} \longrightarrow NaPb(OH)_3$

$CuCl_2 + Cu_{(s)} \longrightarrow 2CuCl_{(s)}$

$2PbO + PbO_2 + 4HNO_3 \longrightarrow 2Pb(NO_3)_2 + 2H_2O + PbO_2$

$2Zn_{(s)} + 2NaOH + 6H_2O_{(l)} \longrightarrow 2NaZn(OH)_4 + 3H_{2(aq)}$

$2Al_{(s)} + 2NaOH + 6H_2O_{(l)} \longrightarrow 2NaAl(OH)_{4(q)} + 3H_{2(aq)}$

$Si_{(s)} + 2NaOH_{(aq)} + H_2O_{(l)} \longrightarrow Na_2SiO_{3(aq)} + 2H_{2(g)}$

$2NaOH_{(s)} + CO_{2(q)} + 9H_2O_{(l)} \longrightarrow Na_2CO_3 + 10H_2O_{(l)}$

$Ca(OH)_2 + Ca(HCO_3)_2 \longrightarrow 2CaCO_3 + 2H_2O$

$Cl_2O_{7(q)} + H_2O_{(l)} \longrightarrow 2HClO_{4(aq)}$

$P_4O_{6(s)} + 6H_2O_{(l)} \longrightarrow 4H_3PO_{3(aq)}$

$HCl_{(aq)} + NH_{3(g)} \longrightarrow NH_4Cl_{(s)}$

$H_{(g)} + H_{(g)} \longrightarrow H_{2(g)}$

$O_{2(g)} + O_{2(g)} \longrightarrow 2O_{2(g)}$

$Cl_{2(g)} + Cl_{2(g)} \longrightarrow 2Cl_{2(g)}$

$FeCl_{2(s)} + \tfrac{1}{2}Cl_{2(g)} \longrightarrow FeCl_{3(s)}$

$Fe_{(s)} + S_{(s)} \longrightarrow FeS_{(s)}$

$PbO_{2(s)} + SO_{2(g)} \longrightarrow PbSO_{4(s)}$

$Ag^{+}_{(aq)} + Cl^{-}_{(aq)} \longrightarrow AgCl_{(s)}$

$H_{2(s)} + I_{2(g)} \longleftrightarrow 2HI_{(g)}$

$N_{2(g)} + 3H_{2(g)} \longleftrightarrow 2NH_{3(g)}$

$C_{(s)} + 2S_{(s)} \longrightarrow CS_{2(l)}$

$CO_{(g)} + 2H_{2(g)} \longleftrightarrow CH_3.OH$

$Ti + 2Cl_2 \longrightarrow TiCl_4$

$TiCl_4 + Ti \longrightarrow 2TiCl_2$

$4OH^- + 4CO_2 \longrightarrow 4HCO_3^-$

$CaO + SiO_2 \longrightarrow CaSiO_3$

$Fe^{2+} + 2OH^- \longrightarrow Fe(OH)_2$

$Fe^{2+} + CO_3^{2-} \longrightarrow FeCO_3$

$Ni + 4CO \xrightarrow{60^0C} Ni(CO)_4$

$Fe + S \longrightarrow FeS$

$2Fe_{(s)} + 3Cl_2 \longrightarrow 2FeCl_{3(s)}$

3. DEHYDROLIZATION (DEHYDRATION) REACTION EQUATIONS

$2H_2O_{2(aq)} \longrightarrow 2H_2O_{(l)} + O_{2(g)}$

$Ca(OH)_{2(aq)} + CO_{2(g)} \longrightarrow CaCO_{3(s)} + H_2O_{(g)}$

$2Al(OH)_3 \longrightarrow Al_2O_{3(s)} + 3H_2O$

$3Zn_{(s)} + 8HNO_{3(q)} \longrightarrow 3Zn(NO_3)_{2(aq)} + 4H_2O + 2NO_{(g)}$

$2Fe(OH)_{3(s)} \longrightarrow Fe_2O_{3(s)} + 3H_2O_{(g)}$

$CUO_{(s)} + H_{2(g)} \longrightarrow CU + H_2O$

$Ca(HCO_3)_2 \longrightarrow CaCO_3 + H_2O + CO_2$

$Ca(OH)_2 + Ca(HCO_3)_2 \longrightarrow 2CaCO_3 + 2H_2O$

$CuSO_4 \cdot 5H_2O \longrightarrow CuSO_4 + 5H_2O$

$Zn(OH)_2 + H_2SO_4 \longrightarrow ZnSO_4 + 2H_2O$

$PbCO_3 + 2HNO_3 \longrightarrow Pb(NO_3)_2 + H_2O + CO_2$

$NH_4OH + HCl \longrightarrow NH_4Cl + H_2O$

$CH_3CH_2OH_{(aq)} + O_{2(g)} \longrightarrow CH_3COOH_{(aq)} + H_2O_{(l)}$

$CO_3^{2-} + 2H^+_{(aq)} \longrightarrow H_2O_{(l)} + CO_{2(g)}$

$HCO_3^-{}_{(s)} + H^+_{(aq)} \longrightarrow CO_{2(g)} + H_2O_{(l)}$

$H_2C_2O_{4(aq)} + NaOH_{(aq)} \longrightarrow NaHC_2O_{4(aq)} + H_2O$

$CuO + H_2SO_4 \longrightarrow CuSO_4 + H_2O$

$C_2H_5OH_{(l)} + 3O_{2(g)} \longrightarrow 2CO_{2(g)} + 3H_2O$

$Ca(OH)_{2(aq)} + 2HNO_{3(aq)} \longrightarrow Ca(NO_3)_{2(aq)} + 2H_2O_{(l)}$

$Pb_3O_4 + 4H_{2(g)} \longrightarrow 3Pb + 4H_2O$

$C_2H_5OH_{(l)} \longrightarrow C_2H_{4(g)} + H_2O_{(g)}$

$2CrO_4^{2-} + 2H^+ \longleftarrow Cr_2O_7^{2-} + H_2O$

$CrO_4Cl_2 + 4OH^- \longrightarrow CrO_4^{2-} + 2Cl^- + 2H_2O$

$(NH_4)_2Cr_2O_7 \longrightarrow Cr_2O_3 + 4H_2O + N_{2(g)}$

$FeSO_4 \cdot 7H_2O \longrightarrow FeSO_4 + 7H_2O$

$2Fe(H_2O)_3(OH)_3 \longrightarrow Fe_2O_3 + 9H_2O$

$2Pb_3O_{4(s)} + 6H_2SO_{4(aq)} \longrightarrow 6PbSO_{4(s)} + 6H_2O_{(l)} + O_{2(g)}$

4. DEOXYGENATION REACTION EQUATIONS

$2ClO_3^-{}_{(s)} \longrightarrow 2Cl^-_{(s)} + 3O_{2(g)}$

$2H_2O_{2(aq)} \longrightarrow 2H_2O_{(l)} + O_{2(g)}$

$2H_2O_{(g)} \longrightarrow 2H_{2(g)} + O_{2(g)}$

$2Na_2O_{2(s)} + 2H_2O \longrightarrow 4NaOH(aq) + O_2(g)$

$2Na_2O_{2(s)} + 2CO_{2(aq)} \longrightarrow 2Na_2CO_{3(s)} + O_{2(g)}$

$2NO_3^-{}_{(aq)} \longrightarrow 2NO_2^- + O_{2(g)}$

$CuO_{(s)} + CO \longrightarrow Cu + CO_2$

$CuO_{(s)} + 2HCl \longrightarrow CuCl_2 + H_2O_{(l)}$

$PbO + CO_{(g)} \longrightarrow Pb + CO_{2(g)}$

5. DISPLACEMENT REACTION EQUATIONS

$CaCl_2 + Na_2CO_3 \longrightarrow CaCO_3 + 2NaCl$

$Pb(NO_3)_2 + 2NaCl \longrightarrow PbCl_2 + 2NaNO_3$

$BaCl_2 + H_2SO_4 \longrightarrow BaSO_4 + 2HCl$

$Pb(C_2H_3O_2)_2 + H_2S \longrightarrow PbS + 2CH_3COOH$

$Ca(OH)_2 + (NH_4)_2SO_4 \longrightarrow CaSO_4 + 2H_2O + 2NH_{3(g)}$

$3Pb_{(s)} + 8HNO_3 \longrightarrow 3Pb(NO_3)_2 + 4H_2O + 2NO_{(g)}$

$Zn(OH)_2 + H_2SO_4 \longrightarrow ZnSO_4 + 2H_2O$

$ZnCO_3 + H_2SO_4 \longrightarrow ZnSO_4 + H_2O + CO_2$

$PbCO_3 + 2HNO_3 \longrightarrow Pb(NO_3)_2 + H_2O + CO_2$

$NH_4OH + HCl \longrightarrow NH_4Cl + H_2O$

$CuO + H_2SO_4 \longrightarrow CuSO_4 + H_2O$

$Zn + H_2SO_4 \longrightarrow ZnSO_4 + H_2$

$2H_2S + SO_2 \longrightarrow 2H_2O + 3S$

$Mg + H_2SO_4 \longrightarrow MgSO_4 + H_2$

$Zn + CuSO_4 \longrightarrow ZnSO_4 + Cu$

$CuSO_{4(aq)} + H_2S_{(g)} \longrightarrow CuS_{(s)} + H_2SO_{4(aq)}$

$CaCO_{3(s)} + 2HCl_{(aq)} \longrightarrow CaCl_{2(s)} + CO_{2(g)} + H_2O_{(g)}$

$Ca(OH)_{2(aq)} + 2HNO_{3(aq)} \longrightarrow Ca(NO_3)_{2(aq)} + 2H_2O_{(l)}$

$Al_2O_{3(s)} + 6HCl_{(aq)} \longrightarrow 2AlCl_{3(aq)} + 3H_2O_{(l)}$

$2NaS + H_2 \longrightarrow 2NaH + 2S$

$SiCl_{4(l)} + 4H_2O_{(l)} \longrightarrow Si(OH)_{4(aq)} + 4HCl_{(aq)}$

$H_2S_{(g)} + PbNO_3 \longrightarrow PbS_{(s)} + 2HNO_{3(aq)}$

$CuO_{(s)} + H_{2(g)} \longrightarrow H_2O + Cu_{(s)}$

$2NH_{3(g)} + 3Cl_{2(g)} \longrightarrow N_{2(g)} + 6HCl_{(g)}$

$3Fe_{(s)} + 4NO_{(g)} \longrightarrow Fe_3O_{4(s)} + 2N_{2(g)}$

$3Fe_{(s)} + 4N_2O_{(g)} \longrightarrow Fe_3O_4 + 4N_{2(g)}$

$2NaOH_{(s)} + CO_{2(g)} \longrightarrow Na_2CO_{3(s)} + H_2O_{(l)}$

$Na_2CO_{3(s)} + H_2SO_{4(aq)} \longrightarrow Na_2SO_{4(aq)} + CO_{2(g)} + H_2O_{(l)}$

$PbO_{(s)} + 2HNO_{3(aq)} \longrightarrow Pb(NO_3)_{2(aq)} + H_2O_{(l)}$

$Zn_{(s)} + 2HCl_{(aq)} \longrightarrow ZnCl_{2(aq)} + H_{2(g)}$

$2NaCl + H_2SO_4 \longrightarrow Na_2SO_4 + 2HCl_{(g)}$

$Na_2B_4O_7 + H_2SO_4 + 5H_2O \longrightarrow Na_2SO_{4(aq)} + 4H_3BO_3$

$Pb(C_2H_3O_2)_2 + H_2S \longrightarrow PbS + 2CH_3COOH$

$Pb(NO_3)_2 + H_2SO_4 \longrightarrow PbSO_4 + 2HNO_3$

$Na_2CO_{3(s)} + 2HCl_{(aq)} \longrightarrow 2NaCl_{(s)} + H_2O_{(l)} + CO_2$

$2NaHCO_{3(aq)} + H_2SO_{4(aq)} \longrightarrow Na_2SO_4 + 2H_2O_{(l)} + 2CO_{2(g)}$

$Al_2S_3 + 6H_2O_{(l)} \longrightarrow 2Al(OH)_3 + 3H_2S_{(g)}$

$Al(OH)_3 + 3HCl_{(aq)} \longrightarrow AlCl_3 + 3H_2O$

$3Fe_{(s)} + 4H_2O_{(g)} \longrightarrow Fe_3O_{4(s)} + 4H_{2(g)}$

$Mg_{(s)} + 2HCl_{(aq)} \longrightarrow MgCl_{2(aq)} + H_{2(g)}$

$2K_{(s)} + 2H_2O_{(l)} \longrightarrow 2KOH_{(aq)} + H_{2(g)}$

$Ag_2O_{(s)} + H_2O_{2(g)} \longrightarrow 2Ag_{(s)} + H_2O + O_{2(g)}$

$KNO_{3(aq)} + H_2SO_{4(aq)} \longrightarrow KHSO_{4(aq)} + HNO_{3(aq)}$

NaOH + HCl $\longrightarrow$ NaCl + H$_2$O

KOH + HNO$_3$ $\longrightarrow$ KNO$_3$ + H$_2$O

TiCl$_4$ + 2Mg $\longrightarrow$ Ti + 2MgCl$_2$

4FeCr$_2$O$_4$ + 8Na$_2$CO$_3$ + 7O$_2$ $\longrightarrow$ 8Na$_2$CrO$_4$ + 2Fe$_2$O$_3$ + 8CO$_2$

Cr$_2$O$_3$ + 2Al $\longrightarrow$ Al$_2$O$_3$ + 2Cr

Cr$_2$O$_7^{2-}$ + 8H$^+$ + 3H$_2$S $\longrightarrow$ 2Cr^{3+} + 7H$_2$O + 3S

2NaCl$_{(aq)}$ + Pb(NO$_3$)$_{2(aq)}$ $\longrightarrow$ PbCl$_{2(s)}$ + 2NaNO$_{3(aq)}$

3Mn$_3$O$_4$ + 8Al $\longrightarrow$ 4Al$_2$O$_{3(l)}$ + 9Mn

2H$_2$S + SO$_2$ $\longrightarrow$ 4H$_2$O$_{(l)}$ + 3S$_{(s)}$

3Zn$_{(s)}$ + 8HNO$_{3(aq)}$ $\longrightarrow$ 3Zn(NO$_3$)$_{2(aq)}$ + 4H$_2$O$_{(l)}$ + 2NO$_{(g)}$

3Cu$_{(s)}$ + 8HNO$_{3(aq)}$ $\longrightarrow$ 3(Cu^{2+})(NO$_3$)$_{2(aq)}$ + 4H$_2$O + 2NO$_{(g)}$

6. ELECTRON TRANSFER EQUATIONS

2H$^+_{(aq)}$ + O$_{3(g)}$ + 2e$^-$ $\longrightarrow$ O$_{2(g)}$ + H$_2$O$_{(l)}$

2I$^-_{(aq)}$ + 2H$^+_{(aq)}$ + O$_{3(g)}$ $\longrightarrow$ I$_{2(aq)}$ + O$_{2(g)}$ + H$_2$O$_{(l)}$

O$^{2-}_{(s)}$ + C$_{(s)}$ $\longrightarrow$ CO$_{(g)}$ + 2e$^-$

Pb$^{2+}_{(s)}$ + 2e$^-$ $\longrightarrow$ Pb$_{(s)}$

Zn $\longrightarrow$ Zn$^{2+}_{(aq)}$ + 2e$^-$

CH$_3$.COO$^-_{(aq)}$ + H$^+$ + Na$^+$ + OH$^-_{(aq)}$ $\longrightarrow$ CH$_3$COONa$_{(aq)}$ + H$_2$O$_{(l)}$

CO$_3^-{}_{(s)}$ + 2H$^+_{(aq)}$ $\longrightarrow$ H$_2$O$_{(l)}$ + CO$_{2(g)}$

Mg$_{(s)}$ + 2H$^+_{(aq)}$ $\longrightarrow$ Mg$^{2+}_{(aq)}$ + H$_{2(g)}$

2Fe$^{2+}_{(aq)}$ + Cl$_{2(g)}$ $\longrightarrow$ 2Fe$^{3+}_{(aq)}$ + 2Cl$^-_{(aq)}$

Cl$_{2(g)}$ + 2Br$^-_{(aq)}$ $\longrightarrow$ 2Cl$^-_{(aq)}$ + Br$_{2(g)}$

Br$_{2(g)}$ + 2I$^-_{(aq)}$ $\longrightarrow$ 2Br$^-_{(aq)}$ + I$_{2(g)}$

2H$^-_{(aq)}$ $\longrightarrow$ H$_{2(g)}$ + 2e$^-$

2Mn$^{4+}_{(aq)}$ + H$_{2(g)}$ + 2OH$^-_{(aq)}$ $\longrightarrow$ 2Mn$^{3+}_{(aq)}$ + 2H$_2$O$_{(l)}$

$Mg_{(s)} + Cl_{2(g)} \longrightarrow Mg^{2+} + 2Cl^{-}_{(s)}$

$Mg_{(s)} + S_{(s)} \longrightarrow Mg^{2+} + S^{2-}_{(s)}$

$Fe^{2+}_{(aq)} + \frac{1}{2}Cl_{2(g)} \longrightarrow Fe^{3+}_{(aq)} + Cl^{-}_{(aq)}$

$S^{2-}_{(s)} + Cl_{2(g)} \longrightarrow S_{(s)} + 2Cl^{-}_{(aq)}$

$6Fe^{2+}_{(aq)} + 8HNO_{3(aq)} \longrightarrow 6Fe^{3+} + 6NO_{3^-(aq)} + 4H_2O_{(l)} + 2NO_{(g)}$

$Ag^{+}_{(ac)} + Cl^{-}_{(aq)} \longrightarrow AgCl_{(s)}$

$2I^{-}_{(aq)} + S_2O_8^{2-} \longrightarrow I_2 + 2SO_4^{2-}$

$2Fe^{3+} + 2I^{-} \longrightarrow 2Fe^{2+} + I_2$

$2Fe^{2+} + S_2O_8^{2-} \longrightarrow 2Fe^{3+} + 2SO_4^{2-}$

$2VO_3^{-} + 4H^{+} + SO_2 \longrightarrow 2VO^{2+} + SO_4^{2-} + 2H_2O$

$2VO_3^{-} + 12H^{+} + 3Zn \longrightarrow 2V^{2+} + 6H_2O + 3Zn^{2+}$

$2Cr^{3+} + 4OH^{-} + 3O_2^{2-} \longrightarrow 2CrO^{2-} + 2H_2O$

$Cr_2O_7^{2-} + 14H^{+} + 6Fe^{2+} \longrightarrow 2Cr^{3+} + 6Fe^{3+} + 7H_2O$

$Cr^{3+} + 3OH^{-} \longrightarrow Cr(OH)_3$

$CrO_2Cl_2 + 4OH^{-} \longrightarrow CrO_4^{2-} + 2Cl^{-} + 2H_2O$

$Cr_2O_7^{2-} + 2HCl \longrightarrow 2CrO_3Cl^{-} + H_2O$

$Cr_2O_3 + 6H^{+} \longleftrightarrow 2Cr^{3+} + 3H_2O$

$2Cr^{3+} + Zn \longrightarrow 2Cr^{2+} + Zn^{2+}$

$3MnO_2 + 6OH^{-} + ClO_3^{-} \longrightarrow 3MnO_4^{2-} + 3H_2O + Cl^{-}$

$3MnO_4^{2-} + 2H_2O \longrightarrow 2MnO_4^{-} + MnO_2 + 4OH^{-}$

$4OH^{-} + 4CO_2 \longrightarrow 4HCO_3^{-}$

$2MnO_4^{2-} + Cl_2 \longrightarrow 2MnO_4^{-} + 2Cl^{-}$

$2Mn^{2+} + 5BiO_3^{-} + 14H^{+} \longrightarrow 2MnO_4^{-} + 5Bi^{3+} + 7H_2O$

$2MnO_4^{-} + 6H^{+} + 5H_2S \longrightarrow 2Mn^{2+} + 8H_2O + 5S$

$MnO_4^{-} + 8H^{+} + 5Fe^{2+} \longrightarrow Mn^{2+} + 5Fe^{3+} + 4H_2O$

$Fe^{2+} + 2OH^- \longrightarrow Fe(OH)_2$

$Co^{2+} + 2OH^- \longrightarrow Co(OH)_2$

$Ni^{2+} + 2OH^- \longleftrightarrow Ni(OH)_2$

$Ti^{2+} + 2e^- \longrightarrow Ti$

$H^+ + e^- \longrightarrow \tfrac{1}{2}H_2$

$Ti + 2H^+ \longrightarrow Ti^{2+} + H_2$

$3MnO_4^{2-} + 4H^+ \longrightarrow 2MnO_4^- + MnO_2 + 2H_2O$

$2Fe^{2+} + 2H^+ + \tfrac{1}{2}O_2 \longrightarrow 2Fe^{3+} + H_2O$

7. HYDROLIZATION REACTION (HYDROLYSIS) EQUATIONS

$P_4O_{10(s)} + 6H_2O_{(l)} \longrightarrow 4H_3PO_{4(aq)}$

$P_4O_{6(s)} + 6H_2O_{(l)} \longrightarrow 4H_3PO_{3(aq)}$

$SO_{2(g)} + H_2O_{(l)} \longrightarrow H_2SO_{3(aq)}$

$CO_{2(g)} + H_2O_{(l)} \longrightarrow H_2CO_{3(aq)}$

$CaO_{(s)} + H_2O_{(l)} \longrightarrow Ca(OH)_{2(aq)}$

$P_4O_{10(s)} + 2H_2O_{(l)} \longrightarrow 4HPO_{3(aq)}$

$ZnO_{(s)} + 2NaOH_{(aq)} + H_2O_{(l)} \longrightarrow Na_2Zn(OH)_4$

$Al_2O_{3(s)} + 2NaOH_{(aq)} + 3H_2O_{(l)} \longrightarrow 2NaAl(OH)_{4(aq)}$

$Na_2O_{(s)} + H_2O_{(l)} \longrightarrow 2NaOH_{(aq)}$

$CaO_{(s)} + H_2O_{(l)} \longrightarrow Ca(OH)_{2(s)}$

$PbO + NaOH + H_2O_{(l)} \longrightarrow NaPb(OH)_3$

$2Na_2O_{2(s)} + 2H_2O_{(l)} \longrightarrow 4NaOH_{(aq)} + O_{2(g)}$

$CH_{4(g)} + 2H_2O_{(g)} \longrightarrow CO_{2(g)} + 4H_{2(g)}$

$CO_{(g)} + 3H_{2(g)} + H_2O_{(g)} \longrightarrow CO_{2(g)} + 4H_{2(g)}$

$2MnO_4^- + 16H^+ + 10H_2O_{(l)} \longrightarrow 2Mn^{2+} + 18H_2O$

$2Zn_{(s)} + 6H_2O_{(l)} + 2NaOH_{(aq)} \longrightarrow 2NaZn(OH)_4 + 3H_2$

$2Al_{(s)} + 2NaOH_{(aq)} + 6H_2O_{(l)} \longrightarrow 2NaAl(OH)_{4(aq)} + 3H_{2(g)}$

$Si_{(s)} + 2NaOH_{(aq)} + H_2O_{(l)} \longrightarrow Na_2SiO_{3(aq)} + 2H_{2(g)}$

$CO_{2(g)} + 2NaOH_{(s)} + 9H_2O_{(l)} \longrightarrow Na_2CO_3 \cdot 10H_2O_{(s)}$

$12FeSO_{4(s)} + 6H_2O_{(l)} + 3O_{2(g)} \longrightarrow 4(Fe_2(SO_4)_3 \cdot Fe(OH)_3)_{(s)}$

$2K_{(s)} + 2H_2O_{(l)} \longleftrightarrow 2KOH_{(aq)} + H_{2(g)}$

$2Na_{(s)} + 2H_2O_{(l)} \longrightarrow 2NaOH_{(aq)} + H_{2(g)}$

$Mg_{(s)} + H_2O \longrightarrow MgO + H_2$

$Zn + H_2O \longrightarrow ZnO + H_2$

$3Fe + 4H_2O \longrightarrow Fe_3O_4 + 4H_2$

$Cl_2 + H_2O \longrightarrow HOCl + HCl$

$K_2O + H_2O \longrightarrow 2KOH$

$2NO_2 + H_2O \longrightarrow HNO_2 + HNO_3$

$CaC_2 + 2H_2O_{(l)} \longrightarrow Ca(OH)_2 + C_2H_2$

$Na_2B_4O_7 + H_2SO_4 + 5H_2O_{(l)} \longrightarrow Na_2SO_{4(aq)} + 4H_3BO_3$

$N_2O_{5(s)} + H_2O_{(l)} \longrightarrow 2HNO_{3(aq)}$

$O^{2-}_{(aq)} + H_2O_{(l)} \longrightarrow 2OH^-_{(aq)}$

$SiCl_{4(l)} + 4H_2O_{(l)} \longrightarrow Si(OH)_{4(aq)} + 4HCl_{(aq)}$

$PCl_{5(l)} + 4H_2O_{(l)} \longrightarrow H_3PO_{4(aq)} + 5HCl_{(aq)}$

$PCl_{3(l)} + 3H_2O_{(l)} \longleftrightarrow H_3PO_{3(aq)} + 3HCl_{(aq)}$

$Cl_2O_{7(g)} + H_2O_{(l)} \longrightarrow 2HClO_{4(aq)}$

$Al_2S_{3(s)} + 6H_2O_{(l)} \longrightarrow 2Al(OH)_3 + 3H_2S_{(g)}$

$C_{(s)} + H_2O_{(l)} \longrightarrow CO_{(g)} + H_{2(g)}$

$2F_{2(g)} + 2H_2O_{(l)} \longrightarrow 4HF_{(aq)} + O_{2(g)}$

$BiCl_{3(aq)} + H_2O_{(l)} \longrightarrow BiOCl_{(s)} + 2HCl_{(aq)}$

$Cu^{2+}_{(s)} + 4H_2O_{(l)} \longleftrightarrow (Cu \cdot 4H_2O)^{2+}_{(ac)}$

$TiCl_4 + 2H_2O \longrightarrow TiO_2 + 4HCl$

$Cr_2O_3 + 6OH^- + 3H_2O \longrightarrow 2Cr(OH)_6^{3-}$

$Mn^{2+} + 2OH^- \longrightarrow Mn(OH)_2$

8. HYDROGENATION REACTION EQUATIONS

$Fe_2O_{3(s)} + 3H_{2(g)} \longrightarrow 2Fe_{(s)} + 3H_2O$

$Fe_3O_{4(s)} + 4H_{2(g)} \longleftrightarrow 3Fe_{(s)} + 4H_2O$

$PbO + H_2 \longrightarrow Pb_{(s)} + H_2O$

$CuO_{(s)} + H_{2(g)} \longrightarrow Cu + H_2O$

$CO_{(g)} + 3H_{2(g)} + H_2O_{(g)} \longrightarrow CO_{2(g)} + 4H_{2(g)}$

$Cl_{2(g)} + H_{2(g)} \longrightarrow 2HCl_{(g)}$

$N_{2(g)} + 3H_{2(g)} \longrightarrow 2NH_{3(g)}$

9. ION EXCHANGE REACTION EQUATIONS

$V_2O_5 + 6OH^- \longrightarrow 2VO_4^{3-} + 3H_2O$

$V_2O_5 + 2H^+ \longleftrightarrow 2VO_4^+ + H_2O$

$CrO_2Cl_2 + 4OH^- \longrightarrow CrO_4^{2-} + 2Cl^- + 2H_2O$

$CrO_3Cl^- + 2OH^- \longleftrightarrow CrO_4^{2-} + Cl^- + H_2O$

$Cr_2O_7^{2-} + 2HCl \longleftrightarrow 2CrO_3Cl^- + H_2O$

$Cr_2O_3 + 6H^+ \longrightarrow 2Cr^{3+} + 3H_2O$

$3MnO_4^{2-} + 4H^+ \longrightarrow 2MnO_4^- + MnO_2 + 2H_2O$

10. IONIZATION REACTION EQUATIONS

$Zn \longrightarrow Zn^{2+}_{(aq)} + 2e^-$

$C_2H_4O_2 \longleftrightarrow H^+ + C_2H_3O_2^-$

$CH_3.COOH \longleftrightarrow CH_3.COO^-_{(aq)} + H^+_{(aq)}$

$2Fe^{2+}_{(aq)} + Cl_{2(g)} \longleftrightarrow 2Fe^{3+}_{(aq)} + 2Cl^-_{(aq)}$

$Br_2 + 2I^-_{(aq)} \longrightarrow 2Br^-_{(aq)} + I_{2(s)}$

$Na_{(s)} + \tfrac{1}{2}Cl_2 \longrightarrow Na^+ + Cl^-_{(s)}$

$NH_4OH_{(aq)} \longleftrightarrow NH_4^+_{(aq)} + OH^-_{(aq)}$

$H_2O_{(g)} \longleftrightarrow H^+_{(aq)} + OH^-_{(aq)}$

$KAg(CN)_{2(aq)} \longrightarrow K^+_{(aq)} + Ag^+_{(aq)} + 2CN^-_{(aq)}$

$Mg_{(s)} + Cl_{2(g)} \longrightarrow (Mg^{2+} + 2Cl^-)_{(s)}$

$Mg_{(s)} + S_{(s)} \longrightarrow Mg^{2+} + S^{2-}_{(s)}$

$H_2S \longrightarrow 2H^+ + S^{2-}$

$PbCl_{2(s)} \longrightarrow Pb^{2+}_{(aq)} + 2Cl^-_{(aq)}$

$Cl_{2(g)} \longrightarrow Cl^-_{(g)} + Cl^-_{(g)}$

$2Fe \longrightarrow 2Fe^{2+} + 4e^-$

11. NEUTRALIZATION REACTION EQUATIONS

$Ca(OH)_{2(aq)} + CO_{2(g)} \longrightarrow CaCO_{3(s)} + H_2O_{(l)}$

$CaO_{(s)} + 2HCl_{(aq)} \longrightarrow CaCl_{2(s)} + H_2O_{(l)}$

$ZnO_{(s)} + H_2SO_{4(aq)} \longrightarrow ZnSO_{4(s)} + H_2O_{(l)}$

$Al_2O_{3(s)} + 6HCl_{(aq)} \longrightarrow 2AlCl_3 + 3H_2O_{(l)}$

$Ca(OH)_2 + 2HCl_{(aq)} \longrightarrow CaCl_{2(s)} + 2H_2O_{(l)}$

$CuO_{(s)} + 2HCl_{(aq)} \longrightarrow CuCl_2 + H_2O_{(l)}$

$PbO_{2(s)} + 4HCl_{(aq)} \longrightarrow PbCl_{2(s)} + 2H_2O + Cl_{2(g)}$

$Pb_3O_4 + 4HNO_{3(aq)} \longrightarrow 2Pb(NO_3)_2 + 2H_2O_{(l)} + PbO_2$

$2PbO + PbO_2 + 4HNO_{3(aq)} \longrightarrow 2Pb(NO_3)_2 + 2H_2O + PbO_2$

$Pb_3O_{4(s)} + 8HCl_{(aq)} \longrightarrow 3PbCl_{2(s)} + 4H_2O + Cl_{2(g)}$

$2Pb_3O_{4(s)} + 6H_2SO_{4(aq)} \longrightarrow 6PbSO_{4(s)} + 6H_2O_{(l)} + O_{2(g)}$

$2KI_{(aq)} + H_2SO_{4(aq)} + O_{3(g)} \longrightarrow I_{2(g)} + O_{2(g)} + K_2SO_{4(aq)} + H_2O_{(l)}$

$BaO_{2(s)} + H_2SO_{4(aq)} \longrightarrow BaSO_{4(s)} + H_2O_{(l)} + \tfrac{1}{2}O_2$

$PbO_{(s)} + 2HNO_{3(aq)} \longrightarrow Pb(NO_3)_{2(aq)} + H_2O_{(l)}$

$Ca(OH)_2 + Ca(HCO_3)_2 \longrightarrow 2CaCO_3 + 2H_2O$

$2NaCl + H_2SO_4 \longleftrightarrow Na_2SO_4 + 2HCl_{(g)}$

$Zn(OH)_2 + H_2SO_4 \longrightarrow ZnSO_4 + 2H_2O$

$PbCO_3 + 2HNO_3 \longrightarrow Pb(NO_3)_2 + H_2O + CO_2$

$NH_4OH + HCl \longrightarrow NH_4Cl + H_2O$

$H_2C_2O_{4(aq)} + NaOH_{(aq)} \longrightarrow NaHC_2O_{4(aq)} + H_2O$

$CH_3.COOH_{(aq)} + C_2H_5OH_{(aq)} \longrightarrow CH_3.COOC_2H_5 + H_2O$

$Ca(OH)_{2(aq)} + 2HNO_{3(aq)} \longrightarrow Ca(NO_3)_{2(aq)} + 2H_2O_{(l)}$

$Al_2O_{3(s)} + 6HCl_{(aq)} \longrightarrow 2AlCl_{3(aq)} + 3H_2O_{(l)}$

$6HCl_{(g)} + 6NH_{3(g)} \longrightarrow 6NH_4Cl_{(s)}$

$8NH_{3(g)} + 3Cl_2 \longrightarrow 6NH_4Cl_{(s)} + N_{2(g)}$

$2NaOH + CO_{2(g)} \longrightarrow Na_2CO_3 + H_2O_{(l)}$

$2KOH_{(aq)} + CO_{2(g)} \longrightarrow K_2CO_{3(s)} + H_2O_{(l)}$

$NaOH_{(aq)} + HCl_{(aq)} \longrightarrow NaCl_{(s)} + H_2O_{(l)}$

$2NaHCO_{3(aq)} + H_2SO_{4(aq)} \longrightarrow Na_2SO_{4(s)} + 2H_2O_{(l)} + 2CO_{2(g)}$

$Na_2CO_{3(aq)} + 2HCl_{(aq)} \longrightarrow 2NaCl_{(aq)}.H_2O_{(l)} + CO_{2(g)}$

$Al(OH)_{3(aq)} + 3HCl_{(aq)} \longrightarrow AlCl_3 + 3H_2O$

$3Cu_{(s)} + 8HNO_{3(aq)} \longrightarrow 3(Cu^{2+}2NO_3^-)_{(aq)} + 4H_2O_{(l)} + 2NO_{(g)}$

$Fe_2(SO_4)_{3(aq)} + 2KI_{(aq)} \longrightarrow 2FeSO_{4(aq)} + K_2SO_{4(g)} + I_{2(g)}$

$2KMnO_4 + 5H_2O + 3H_2SO_4 \longrightarrow K_2SO_4 + 2MnSO_4 + 2½O_2 + 8H_2O$

$KOH + HNO_3 \longrightarrow KNO_3 + H_2O$

$2CrO_4^{2-} + 2H^+ \longleftrightarrow Cr_2O_7^{2-} + H_2O$

$Cr_2O_7^{2-} + 8H^+ + 3H_2S \longrightarrow 2Cr^{3+} + 7H_2O + 3S$

$Cr_2O_7^{2-} + 14H^+ + 6Fe^{2+} \longrightarrow 2Cr^{3+} + 6Fe^{3+} + 7H_2O$

$Cr_2O_7^{2-} + 2HCl \longrightarrow 2CrO_3Cl^- + H_2O$

$Cr_2O_3 + 6H^+ \longrightarrow 2Cr^{3+} + 3H_2O$

$2MnO_4^- + 16H^+ + 5Zn \longrightarrow 2Mn^{2+} + 5Zn^{2+} + 8H_2O$

$2MnO_4^- + 6H^+ + 5H_2S \longrightarrow 2Mn^{2+} + 8H_2O + 5S$

$MnO_4^- + 8H^+ + 5Fe^{2+} \longrightarrow Mn^{2+} + 5Fe^{3+} + 4H_2O$

$2MnO_4^- + 16H^+ + 5C_2O_4^{2-} \longrightarrow 2Mn^{2+} + 10CO_2 + 8H_2O$

$MnO_2 + 4HCl \longrightarrow MnCl_2 + 2H_2O + Cl_2$

$2FeSO_4 + H_2SO_3 + H_2O \longrightarrow Fe_2(SO_4)_3 + 2H_2O$

$Au_2O_{3(s)} + 6HNO_{3(aq)} \longrightarrow Au_2(NO_3)_{6(aq)} + 3H_2O_{(l)}$

12. OXYGENATION REACTION EQUATIONS

$2NO_{(g)} + O_2 \longrightarrow 2NO_{2(g)}$

$2Mg_{(s)} + O_{2(g)} \longrightarrow 2MgO_{(s)}$

$P_{4(s)} + 5O_{2(s)} \longrightarrow P_4O_{10(s)}$

$P_{4(g)} + 3O_{2(g)} \longrightarrow P_4O_{6(s)}$

$S_{(s)} + O_{2(g)} \longrightarrow SO_{2(g)}$

$2C + O_2 \longrightarrow 2CO$

$C_{(s)} + O_{2(g)} \longrightarrow CO_{2(g)}$

$2Zn_{(s)} + O_{2(g)} \longrightarrow 2ZnO_{(aq)}$

$2Na_{(s)} + O_{2(g)} \longrightarrow Na_2O_{2(s)}$

$6PbO + O_2 \longrightarrow 2Pb_3O_{4(s)}$

$3O_{2(g)} \longrightarrow 2O_{3(g)}$

$PbS_{(s)} + 4O_{3(g)} \longrightarrow PbSO_{4(s)} + 4O_{2(g)}$

$H_2S_{(g)} + 4O_{3(g)} \longrightarrow H_2SO_{4(aq)} + 4O_{2(g)}$

$2I^-_{(aq)} + 2H^+ + O_{3(g)} \longrightarrow I_{2(aq)} + O_{2(g)} + H_2O_{(l)}$

$PbS_{(s)} + 4H_2O_{2(aq)} \longrightarrow PbSO_{4(s)} + 4H_2O$

$O_{3(g)} + H_2O_{2(aq)} \longrightarrow H_2O_{(l)} + 2O_{2(g)}$

$2CO_{(g)} + O_{2(g)} \longrightarrow 2CO_{2(g)}$

$2Cu_{(s)} + O_{2(g)} \longrightarrow 2CuO$

$2Pb_{(s)} + O_{2(g)} \longrightarrow 2PbO_{(s)}$

$2Mg_{(s)} + O_{2(g)} \longrightarrow 2MgO_{(s)}$

$C_6H_{12}O_{6(s)} + 6O_{2(g)} \longrightarrow 6CO_{2(g)} + 6H_2O_{(l)}$

$CH_3.CHO_{(aq)} + \tfrac{1}{2}O_{2(g)} \longrightarrow CH_3.COOH$

$CH_3.CH_2OH_{(aq)} + O_{2(g)} \longrightarrow CH_3COOH_{(aq)} + H_2O$

$H_2O_{(l)} + O^{2-}_{(aq)} \longrightarrow 2OH^-$

$2CH_{4(g)} + 4O_{2(g)} \longrightarrow 2CO_{2(g)} + 4H_2O_{(g)}$

$2H_{2(g)} + O_{2(g)} \longrightarrow 2H_2O_{(g)}$

$O_{2(g)} + O_{2(g)} \longrightarrow 2O_{2(g)}$

$N_{2(g)} + 2O_{2(g)} \longleftrightarrow 2NO_2(g)$

$N_{2(g)} + \tfrac{1}{2}O_{2(g)} \longleftrightarrow N_2O_{(g)}$

$N_{2(g)} + O_{2(g)} \longleftrightarrow 2NO_{(g)}$

$2SO_{2(g)} + O_{2(g)} \longleftrightarrow 2SO_{3(g)}$

$C_{12}H_{22}O_{11(s)} + 12O_{2(g)} \longrightarrow 12CO_{2(g)} + 11H_2O_{(l)}$

13. REDOX REACTION EQUATIONS

$2Na_2O_2 + 2CO_{2(aq)} \longrightarrow 2Na_2CO_{3(g)} + O_{2(g)}$

$PbO_{(s)} + 2H_2O_{2(aq)} \longrightarrow PbO_{(s)} + 2H_2O_{(s)} + O_{2(g)}$

$Zn_{(s)} + 2HCl_{(aq)} \longrightarrow ZnCl_{2(aq)} + H_{2(g)}$

$Na_2CO_3 + Ca(HCO_3)_2 \longrightarrow 2NaHCO_3 + CaCO_3$

$Ca(HCO_3)_2 + 2Na_2St \longrightarrow CaSt_2 + 2NaHCO_3$

$CaSO_4 + Na_2CO_3 \longrightarrow CaCO_3 + Na_2SO_4$

$CaY + 2NaCl \longrightarrow Na_2Y + CaCl_2$

$Pb(C_2H_3O_2)_2 + H_2S \longrightarrow PbS + 2CH_3COOH$

$Ca(OH)_2 + (NH_3)SO_4 + H_{2(g)} \longrightarrow CaSO_4 + 2H_2O + NH_{3(g)}$

$Pb(NO_3)_2 + H_2SO_4 \longrightarrow PbSO_4 + 2HNO_3$

$BaCl_2 + H_2SO_4 \longrightarrow BaSO_4 + 2HCl$

$Pb(NO_3)_2 + 2NaCl \longrightarrow PbCl_2 + 2NaNO_3$

$Fe + S \longrightarrow FeS$

$2Fe_{(s)} + 3Cl_2 \longrightarrow 2FeCl_2$

$CuSO_{4(aq)} + H_2S_{(g)} \longrightarrow CuS + H_2SO_{4(aq)}$

$2NaS + H_{2(g)} \longrightarrow 2NaH + 2S$

$2NH_{3(g)} + 3Cl_{2(g)} \longrightarrow N_{2(g)} + 6HCl_{(g)}$

$V_2O_5 + 2H_2 \longrightarrow V_2O_3 + 2H_2O$

$NH_4Cl_{(s)} + NaOH_{(aq)} \longrightarrow NaCl_{(aq)} + NH_{3(g)} + H_2O_{(l)}$

$H_2O_{(g)} + C_{(s)} \longrightarrow H_{2(g)} + CO_{(g)}$

$SO_{2(g)} + H_2O + NaClO \longrightarrow NaCl_{(s)} + H_2SO_{4(aq)}$

$2FeCl_{2(s)} + Cl_{2(g)} \longrightarrow 2FeCl_{3(s)}$

$Ag_2O_{(s)} + H_2O_{2(g)} \longrightarrow 2Ag_{(s)} + H_2O + O_{2(g)}$

$Ba(OH)_{2(aq)} + H_2O_{2(g)} \longrightarrow BaO_2 + 2H_2O_{(l)}$

$KNO_{3(aq)} + H_2SO_{4(aq)} \longrightarrow KHSO_{4(aq)} + HNO_{3(aq)}$

$3VO_4^{3-} + 6H^+ \longleftrightarrow V_3O_9^{3-} + 3H_2O$

$V_2O_5 + 6H^+ + 2e^- \longrightarrow 2VO^{2+} + 3H_2O$

$FeCr_2O_4 + 4C \longrightarrow Fe + 2Cr + 4CO$

$4FeCr_2O_4 + 8Na_2CO_3 + 7O_2 \longrightarrow 8Na_2CrO_4 + 2Fe_2O_3 + 8CO_2$

$2Na_2CrO_4 + H_2SO_4 \longrightarrow Na_2Cr_2O_7 + Na_2SO_4 + H_2O$

$Na_2Cr_2O_7 + 2C \longrightarrow Cr_2O_3 Na_2CO_3 + CO$

$CrO_3Cl^- + 2OH^- \longrightarrow CrO_4^{2-} + Cl^- + H_2O$

$Cr^{3+} + 3OH^- \longrightarrow Cr(OH)_3$

$K_2Cr_2O_7 + H_2SO_4 + 3SO_2 \longrightarrow K_2SO_4 + Cr_2(SO_4)_3 + H_2O$

$SiO_2 + 2C \longrightarrow Si + 2CO$

$Fe_2O_3 + 3C \longrightarrow 2Fe + 3CO$

$FeS + 2H^+ \longrightarrow Fe^{2+} + H_2S$

$NiO + C \longrightarrow Ni + CO$

$NiO + CO \longrightarrow Ni + CO_2$

$Fe + CO + O_2 \longrightarrow FeCO_3$

14. REVERSIBLE REACTION EQUATIONS

$2Pb_3O_{4(s)} \longleftrightarrow 6PbO_{(s)} + O_{2(g)}$

$Cl_2 + H_2O \longleftrightarrow HOCl + HCl$

$C_2H_4O_4 \longleftrightarrow H^+ + C_2H_3O_4^-$

$CH_3.COOH \longleftrightarrow CH_3.COO^-_{(aq)} + H^+_{(aq)}$

$CH_3.COOH_{(aq)} + C_2H_5OH_{(l)} \longleftrightarrow CH_3.COOC_2H_5 + H_2O$

$NH_4OH_{(aq)} \longleftrightarrow NH_4^+_{(aq)} + OH^-_{(aq)}$

$H_2O_{(l)} \longleftrightarrow H^+_{(aq)} + OH^-_{(aq)}$

$KAg(CN)_{2(aq)} \longleftrightarrow K^+_{(aq)} + Ag^+_{(aq)} + 2CN^-_{(aq)}$

$H_2S \longleftrightarrow 2H^+ + S^{2-}$

$PbCl_{2(s)} \longleftrightarrow Pb^{2+}_{(aq)} + 2Cl^-_{(aq)}$

$H_{2(s)} + I_{2(g)} \longleftrightarrow 2HI_{(g)}$

$H_2O_{(l)} \longleftrightarrow H_2O_{(g)}$

$N_{2(g)} + 2O_{2(g)} \longleftrightarrow 2NO_{2(g)}$

$N_{2(g)} + O_{2(g)} \longleftrightarrow 2NO_{(g)}$

$BiCl_{3(aq)} + H_2O_{(l)} \longleftrightarrow BiOCl_{(s)} + 2HCl_{(aq)}$

$N_{2(g)} + 2H_{2(g)} \longleftrightarrow 2NH_{3(g)}$

$2SO_{2(g)} + 2O_{2(g)} \longleftrightarrow 2SO_{4(g)}$

$NH_4Cl_{(s)} \longleftrightarrow NH_{3(g)} + HCl_{(g)}$

$CO + 2H_2 \longleftrightarrow CH_3.OH$

$2CrO_4^{2-} + 2H^+ \longleftrightarrow Cr_2O_7^{2-} + H_2O$

$3MnO_4^{2-} + 2H_2O \longleftrightarrow 2MnO_4^- + MnO_2 + 4OH^-$

$3MnO_4^{2-} + 2H_2 \longleftrightarrow 2MnO_4^- + MnO_2 + 4OH^-$

$Fe_2O_3 + 3CO \longleftrightarrow 2Fe + 3CO_2$

$[Co(H_2O)_6]^{2+} + 4Cl^- \longleftrightarrow [CoCl_4]^{2+} + 6H_2O$

$[Fe(H_2O)_6]^{3+} + H_2O \longleftrightarrow [Fe(H_2O)_5(OH)]^{2+} + H_3O^+$

$[Fe(H_2O)_5(OH)]^{2+} + H_2O \longleftrightarrow [Fe(H_2O)_4(OH)_2]^+ + H_3O^+$

$[Ti(H_2O)_x]^{4+} + 2H_2O \longleftrightarrow [TiO(H_2O)_{x-1}]^{2+} + 2H_3O^+$

15. SUBSTITUTION REACTION EQUATIONS

$Al_2(SO_4)_3 + 6NH_4OH_{(aq)} \longrightarrow 2Al(OH)_{3(aq)} + 3(NH_4)_2SO_4$

$2NaCl + H_2SO_4 \longrightarrow Na_2SO_4 + 2HCl_{(g)}$

$2NaCl_{(s)} + Pb(NO_3)_{2(aq)} \longrightarrow PbCl_{2(s)} + 2NaNO_{3(aq)}$

$2NaS + H_{2(g)} \longrightarrow 2NaH_{(s)} + 2S_{(s)}$

$2KOH + CO_{2(g)} \longrightarrow K_2CO_3 + H_2O_{(l)}$

$2KBr_{(aq)} + Cl_{2(g)} \longrightarrow 2KCl_{(aq)} + Br_{2(l)}$

$[Co(H_2O)_6]^{2+} + 4Cl^- \longrightarrow [CoCl_4]^{2+} + 6H_2O$

$NiO + C \longrightarrow Ni + CO$

$NiO + CO \longrightarrow Ni + CO_2$

16. NUCLEAR REACTION EQUATIONS

$$^{32}_{16}S + ^{1}_{0}n \longrightarrow ^{31}_{15}P + ^{1}_{1}P$$

$$^{210}_{82}Pb \longrightarrow ^{210}_{83}Bi + ^{0}_{-1}e$$

$$^{238}_{92}U \longrightarrow ^{234}_{90}Th + ^{4}_{2}He$$

$$^{14}_{7}N + ^{4}_{2}He \longrightarrow ^{17}_{8}O + ^{1}_{1}P$$

$$^{2}_{1}D + ^{3}_{1}T \longrightarrow ^{4}_{2}He + ^{1}_{0}n$$

$$^{234}_{90}Th \longrightarrow ^{234}_{91}Pa + ^{0}_{-1}e$$

$$^{6}_{3}Li + ^{1}_{0}n \longrightarrow ^{3}_{1}T + ^{4}_{2}He$$

$$^{238}_{92}U + ^{1}_{0}n \longrightarrow ^{239}_{92}U$$

$$^{239}_{92}U \longrightarrow ^{239}_{93}Np + ^{0}_{-1}e$$

$$^{239}_{93}Np \longrightarrow ^{239}_{94}PU + ^{0}_{-1}e$$

$$^{23}_{11}Na + ^{1}_{0}n \longrightarrow ^{24}_{11}Na$$

$$^{59}_{27}Co + ^{1}_{0}n \longrightarrow ^{60}_{27}Co$$

$$^{226}_{88}Ra \longrightarrow ^{222}_{86}Rn + ^{4}_{2}He$$

REFERENCES

1. G.F. Liptrot : Modern Inorganic Chemistry. ELBS ISBN 026306221.X 1971

2. Therald Moeller: Inorganic Chemistry An Advance text book ISBN 52-7487 1965

3. G. F. Liptrot : Inorganic Chemistry Through Experiments ISBN 0263515915 1975

4. C. A. Nwadinigwe: Inorg. Chem. Guide to IUPAC Nomenclature ISBN 978 156 198 X 1985

5. J. D. Lee: Concise Inorganic Chemistry S ISBN 0-44-30179-0 1979

6. A. P. Kreshkov: A course of Analytical Chemistry Vol. 2

7. A. A. Yaroslavtsev: Quantitative Analysis 1977

8. S. T. Bajah & A. Godman: Chemistry: A New Certificate Approach ISBN 0-582-60686-1 1976

9. A. Holderness & J. Lambert: A New Certificate Chemistry HEB ISBN 435-64425-4 1978

10. I.E. Furmer & V. N. Zaitsev: General Chemical Engineering.

    Translated into English by Yu. N. Vereshchagin: MIR Publishers Moscow 1985.

11. Charles E. Mortimer: Chemistry A Conceptual Approach ISBN 0-442-25554-3

12. S. T. Bajah & H. C. O. Aniodoh: Understanding IUPAC Nomenclature In Chemistry

    ISBN 978-129-128-1 1984

12. IUPAC Reports, J.Amer: Chem. Soc. 2, 5523 (1960)

13 Microsoft ® Encarta ® 2008. © 1993-2007 Microsoft Corporation.

ABOUT THE AUTHOR

 'Remi Tijani, a trained educationist was born in Saki town of Oyo state in Nigeria. He had his primary education at L A Primary School, Isia, in Saki and his secondary education at ESS New-Bussa in Niger state from where he proceeded to the University of Ibadan, Nigeria. 'Remi obtained a bachelor degree in Education, Physics and Chemistry (B.ED Phy/Chem) and later had a master degree in Educational Technology.

'Remi Tijani has taught Physics and Chemistry in some Federal Unity Colleges in Nigeria in the last twenty three years with special interest in teaching the practical aspects of the subjects. A former Physics practical examiner with West African examination council, he is presently a federal inspector of education in Nigeria, with great field experience. The accumulated experience over the years resulted in the present book titled: **THE LANGUAGE OF CHEMISTRY**.

Other works by 'Remi Tijani includes:

[1] *"HAT-The chemistry game"*,

[2] *"My Physics practical note"*

[3] *Music of Physics and*

[4] *Physics mental exercise.*

www.ingramcontent.com/pod-product-compliance
Lightning Source LLC
Chambersburg PA
CBHW081727170526
45167CB00009B/3736